Introductory Applied Statistics

Bruce Blaine

Introductory Applied Statistics

With Resampling Methods & R

 Springer

Bruce Blaine
Statistics Program
University of Rochester
Rochester, NY, USA

ISBN 978-3-031-27743-6   ISBN 978-3-031-27741-2   (eBook)
https://doi.org/10.1007/978-3-031-27741-2

This Springer imprint is published by the registered company Springer Nature Switzerland AG
The registered company address is: Gewerbestrasse 11, 6330 Cham, Switzerland

To Patti and Kate

Preface

This is a book, *I hope*, would gain the approval of John Tukey (1915–2000). I certainly have written it in ways that try to honor his legacy to statistics. John Tukey was an academic and consulting statistician of the highest stripe and a towering intellectual figure. Over his productive career, he developed a long list of innovative statistical tests and concepts, most of which are now indispensable tools for the practicing statistician. In 1962 Tukey published a paper, On the Future of Data Analysis in the *Annals of Mathematical Statistics*, which I can only imagine shocked the sensibilities of the academic establishment. In that paper, he essentially claimed that data analysis, rather than statistics, was his field of interest and practice. His paper took academic statistics to task for being too reliant on mathematical epistemologies and models, and too isolated from the sciences—including the particular data, problems, and quantitative methods of the sciences—to be of much practical use. Tukey also urged his field to develop robust statistical methods, to be as interested in exploring data as in using data to confirm hypotheses, and to harness the emerging power of computing for data analysis.

This book introduces students to the field that Tukey envisioned and inspired—a field that applies statistics through data analysis to disciplinary questions, is in dialogue with data for both confirmatory and exploratory purposes, uses statistical methods that are informed by the data and not enslaved to assumptions, and finally, integrates statistical computing skills into the learning and using of statistics.

The key features of this book include:

- *Statistics and statistical inference taught in a data analytic context.* Statistics and quantitative methods are tools for interacting with data. We have specific questions to test in data, but data also has insights to suggest to us, if we are open to them, and we need statistical methods that are suitable for both purposes. It also means that students learn some basic data wrangling tasks in the process.
- *Robust descriptive statistics.* The book strives to give students a broad portfolio of descriptive statistics for summarizing and exploring univariate and bivariate data, including robust descriptive statistics and effect size statistics.

- *Statistics taught within a study design framework.* This book doesn't organize chapters by statistical method (e.g., a t-test chapter), as is common in introductory statistics books. Rather, descriptive and inferential statistics are organized around the type of study design that produced the data. The framework includes four common bivariate models:

  - ANOVA model (categorical X/numeric Y)
  - Proportions model (categorical X/categorical Y)
  - Regression model (numeric X/numeric Y)
  - Logistic model (numeric X/categorical Y)

- To properly and fully understand statistical inference (and not merely its null hypothesis testing component), students of statistics must learn how study design elements like random sampling, manipulated versus observed predictor variables, and control of alternative explanations affect the interpretation of and conclusions about statistical procedures.
- *Statistical inference taught with resampling methods.* You can draw a straight line between Tukey's vision for statistics and modern simulation-based statistical inference. In this book, students learn and use randomization (aka permutation) and bootstrapping methods for hypothesis testing and parameter estimation. Resampling methods are intuitive and, once understood, apply logically and consistently to inferential problems across study design type. Resampling methods also free us from the limitations of theoretical probability distributions, particularly with regard to parametric assumptions. Finally, the computing skills needed to use randomization and bootstrapping are within the reach of the introductory statistics or data science student.
- *Statistical computing with R.* This book incorporates data analytic examples in R into each chapter, and assumes no prior experience with R in students. Files containing R functions and code are provided for each chapter.

Expectations for Student and Instructor

This book is written for use in undergraduate introductory statistics or applied statistics courses. Indeed, the book emerges from lessons and material I teach in my own Applied Statistics with R course. What do I mean by "applied?" It's an approach to learning statistics that is data analytic, emphasizing how statistics are used to address research questions with various kinds of data and how to do that data analysis. It assumes no mathematical background beyond algebra.

The book assumes that students will use R and RStudio on a regular basis, and have access to those programs outside of class for homework and assignments. No prior computing or programming experience is assumed in students, but developing students' statistical computing skills in R is assumed to be a learning goal of the course, so course time should be allotted for that. RStudio Cloud (aka Posit Cloud) is an excellent environment to work in: it is widely accessible outside of class and doesn't require local installs and updates. For instructors wanting students to use R Markdown for assignments, RStudio Cloud also meshes well with R Markdown.

For instructors, some prior experience with the R language and environment is necessary, but your R skills need not be deep or extensive, especially if you have colleagues to whom you can turn with questions and support. If you are new to R or are just learning, teaching with R is the best way to learn the language. Students need support too, and that can be provided in a variety of ways, both in and outside of class.

Chapter Teaching and Learning Goals

Chapter 1 serves two purposes: 1. Get students at the statistics "workbench" immediately, so they can interact with data and R from the first sessions in the course, and 2. Introduce foundational statistics and statistical concepts and vocabulary that can be built upon in subsequent chapters. This is done largely within the context of summarizing a numeric variable, although late in the chapter I do address how we summarize a categorical variable.

Chapter 2 introduces the statistical modeling framework and study design elements that inform data analysis and interpretations throughout the chapters. I introduce the general linear model, how different combinations of X and Y result in different models, how a statistical model guides analysis, what statistical lenses we can/should use to examine a particular question, how we explore data and allow data to teach us, and how sampling and study design elements figure into data analysis and interpretation.

Chapters 3, 4, 5, and 6 starts with the ANOVA model (Chap. 3) because it is a common and intuitive model for beginning students, and the mean difference is a friendly starting point for learning about effect size statistics. Also, some concepts are foundational and re-purposed in subsequent models (e.g., the mean difference -> risk difference). In Chapter 3, I want to teach students that a statistic is a lens through which one can summarize the effect of a treatment variable, with lots of interesting aspects of a group difference to explore.

Chapter 4 is organized much like Chap. 3 and introduces the proportions model, as well as foundational concepts that show up again in the logistic model: probabilities, proportions, and odds—and the statistics built with them.

Chapter 5 moves from group differences (Chaps. 3 and 4) as the quantity of interest—and the statistics with which we summarize those differences—to linear relationships (Chaps. 5 and 6). Here I teach the basics of least squares simple linear regression but maintain the connection of those particular methods to the analytic framework (relationship summary and exploration, robust/nonrobust statistics, study design elements, and interpretation).

Chapter 6 extends some of the statistical concepts from Chapt. 4, and also builds on least squares regression ideas to teach logistic simple regression. The structure and pedagogy of the chapter reflects the previous 3.

Chapters 3, 4, 5, and 6 form a coherent unit that might be called "statistics and data analysis in bivariate models." At this point, it would be reasonable to do some formal assessment of knowledge, via testing and/or a larger assignment or mini-project that requires students to use these methods and write up a data analytic report.

Chapter 7 introduces statistical inference and focuses on null hypothesis testing methods. It is important for students to learn statistical tools and data analytic skills for descriptive (both confirmatory/summary and exploratory) analysis, without the clutter and complications of establishing "statistical significance." The idea is to help students realize that much of data analysis is descriptive, and we learn a lot about an X-Y relationship through descriptive tools. In this chapter, students learn the logic and principles of hypothesis testing, randomization testing as a resampling method, and how to interpret a p-value.

Chapter 8 focuses on parameter estimation and closes the loop on the effect size statistics covered in Chapters 3, 4, 5, and 6, in that statistics are estimates of some "true" relationship size (e.g., mean difference in the population) and our ultimate interest is, in fact, that parameter. The logic and methods of bootstrapping as a second resampling method are taught, along with two bootstrapped confidence intervals (t interval w/bootstrapped standard error, percentile interval). This chapter also addresses factors that affect interval estimation.

Chapters 7 and 8 introduce resampling methods for inference, but the examples are simple. Chapter 9 applies those methods to each of the models covered in Chaps. 3, 4, 5, and 6, with full data analytic examples. Students therefore learn covering how randomization and bootstrapping are applied to the particularities of each statistical model.

Chapters 7, 8 and 9 form a unit that might be called "statistical inference in bivariate models." At this point, it would be reasonable to again do some formal assessment of knowledge, via testing and/or a larger assignment. Additionally, students are now prepared to do a data analytic project, the parameters of which can be freed or constrained to suit the instructor and circumstances.

Chapter 10 covers descriptive and inferential analysis in repeated-measures data, and focuses on the simplest form of a repeated-measures design—the pre-post study. A pre-post design is very commonly used by researchers, and often is the type of study a student will come up with when given the opportunity to design a study of their own to test some hypothesis. Pre-post designs and data, however, have statistical and inferential issues that set them apart from the designs covered in Chaps. 3, 4, 5, and 6 and 9, and thus this material is somewhat set aside in Chap. 10 and can be thought of as additional material.

Rochester, NY, USA   Bruce Blaine

Contents

Chapter 1
Foundations I: Introductory Data Analysis with R

Covered in This Chapter
- Statistics and plots for summarizing a numeric variable
- Statistics and plots for summarizing a categorical variable
- Data analytic methods and R functions for univariate descriptive analysis

In disciplinary research—from Anthropology to Zoology (and every discipline in between!)—studies produce data from multiple variables. Most research questions are bivariate, and as data analysts, we are ultimately interested in exploring the *relationship* between variables, or *effect* of one variable on another. But data analysis always begins with using statistics and plots to summarize, describe, and display data from single variables, or univariate data analysis. Plus, this is an ideal way to introduce and learn R.

1.1 Goals of Data Analysis

What are we looking for in, or trying to do with, data from single variables? Two things—first, we need to **summarize data** for communication and reporting, and second, we need to **explore** data for what it might teach us. Let's dig a little deeper on each of those goals.

Summarize Data Imagine you have the following scores from ten students who took a 10-point quiz: 6, 7, 7, 8, 8, 8, 9, 9, 10, and 10. If someone asked you to describe the data, you *could* say something like "well, there was one 6, two 7s, three 8s, two 9s, and two 10s." While that description is accurate, it is not a summary—it is simply a list of all the participants and their scores. There are two reasons we want to avoid that. First, participant-level data is usually too granular, and noisy, to see

Table 1.1 R functions for descriptive analysis with a numeric variable

Distributional property	Statistics	Plots	R functions
Location What are the typical or most common values?	Mean Trimmed mean Median Mode	Boxplot Histogram Density plot	mean(x) mean(x, tr=) median(x) boxplot(x) hist(x) plot(density(x))
Variability How much do values vary, typically?	Variance Standard deviation IQR MAD		var(x) sd(x) IQR(x) mad(x)
Influential points What values have the potential to disproportionately influence summary statistics?	Boxplot-defined outliers		boxplot(x)$out

trends and patterns in the data. Statistics and plots are tools that reduce the granularity of participant-level data and help us capture important properties of the variable's **distribution**, which is a fancy word for how the values spread out across the scale of measurement. The statistics and plots we learn in this chapter vary in what property of a distribution they capture, and how they do so, so it is important that we know how those statistics work, and what their strengths and limitations are. Second, participant-level data is cumbersome to report, especially when you have many participants in a study. Statistical and graphical summaries contribute to more effective communication and reporting of data analytic results.

Explore Data   Aside from descriptive and reporting responsibilities, we need to be open to what the data can tell us about trends, patterns, and other higher-level outcomes. Too often, data analysts get answers to their specific questions from the data (e.g., Is X positively associated with Y?) but fail to learn from the data (e.g., What does the data show me about the X-Y relationship in general?). Research questions are important, but can also be biased, and they don't anticipate what the data might teach us apart from answers to those questions. To explore data we need to have a range of statistical and graphical tools through which the data can reveal features to us, and we need to use those tools with an attitude of openness and curiosity. The examples in this chapter will try to model exploratory data analysis (Table 1.1).

1.2   Statistics to Summarize a Numeric Variable

In this section we learn statistics for summarizing and exploring a numeric variable, which is a variable measured on a numeric scale (e.g., height, GPA, income). The statistical summary of a numeric variable is guided by the twin goals of summary and exploration. We are interested in four characteristics of a numeric variable's distribution: **location** concerns typical or "average" values, **variability** concerns how much values spread out around some location statistics, and **modality** captures aspects of the shape of a distribution that can reveal subgroupings of scores in our data. Finally, we are interested in knowing if there are any extremely small or large values in our variable's data. As we will see, extreme value can influence summary statistics. The next section covers location and variability statistics in concept and through their formulas. Although we use a computer to calculate statistics throughout this book, formulas help us understand statistics at a conceptual level. If we understand a statistic conceptually, we are more likely to interpret it correctly. The value of plots (i.e., graphical tools) for summary and exploration will be demonstrated in Sect. 1.4.

1.2.1 Location Statistics

Location (also called central tendency) **statistics** capture the value or range of values that are most typical or common in the distribution of values. In other words, where are most people or the typical person *located* along the range of values of the variable? Let's look at four basic location statistics and learn how they summarize the central or typical value of a numeric variable.

- The **mean** is the average of all n values, where n is the sample size (see Formula 1.1). The mean is the geometric balance point in the distribution, or the value that balances the "weight" of the points above and below it. The mean incorporates every value in its calculation, so each value has a "say" in the mean, which is another way of saying that the mean is a sufficient statistic. Two qualities apply to sufficient statistics. First, the mean can be influenced by extremely large or small values. The potential influence of any particular extreme value, however, is moderated by the sample size. The larger n is, the less influence extreme values have over the mean. Second, the mean will change if observations are added to or subtracted from the sample.

$$\bar{x} = \frac{\sum x}{n}$$

(1.1)

- A **trimmed mean** is an adjusted mean. It is the mean of the remaining values after a set percentage is trimmed from the upper and lower ends of the set of ordered values. For example, a 10% trimmed mean would first remove the highest 10% and the lowest 10% of the ordered values and find the mean of the rest (i.e., the middle 80% of the original values). The trimmed mean is less sufficient than the mean, because it excludes some of the data from the calculation, but it is also more resistant than the mean. A statistic is resistant if the presence of extreme values in the data don't affect it much or at all.

- The **median** is the middle value (if n is odd) or the average of the two middle values (if n is even) of the set of ordered x values (see Formulas 1.2a and 1.2b). The median can also be thought of as the 2nd quartile (Q_2, or the 50th percentile) value, or even as a 50% trimmed mean. In contrast to the mean, the median describes location based on one or two values, making the median less sufficient than the mean. However, the median is a more resistant statistic than the mean because the presence of extremely low or high values will not change the middle values.

$$med(x) = x_{(n+1)/2}, \quad \text{if } n \text{ is odd} \tag{1.2a}$$

$$med(x) = \frac{x_{n/2} + x_{(n/2)+1}}{2}, \quad \text{if } n \text{ is even} \tag{1.2b}$$

- The **mode** is the most frequently occurring value. The mode is not a calculated statistic; it is identified from a frequency table. The mode lacks sufficiency because it is based solely on the most frequent value, and lacks resistance because minor changes to the sample values can result in the modal value changing. When a dataset has multiple values with the same or nearly the same high frequency, there are multiple modes. When multiple modes exist, where these modal values are in relation to each other is an important consideration in data analysis. For example, say we ask 10 students how many times they work out per week, and the responses are 2, 3, 4, 4, 4, 5, 5, 5, 6, and 7. Here there are two modes (4 and 5), but their values are very close. If the responses were 1, 2, 2, 2, 3, 4, 5, 7, 7, and 7, we would have two modes (2 and 7) whose values differ by a lot, in terms of exercise frequency per week. Although both of these datasets are multimodal, the second dataset suggests that we may have two distinctly different subgroups of students in our sample—one that works out much more than the other. Multimodality can therefore be an indicator of the presence of subgroups in our data that differ in the variable being measured.

Due to their high resistance, the trimmed mean and median are examples of **robust statistics**, which is a category of statistics that are less influenced by extreme values. The distinction between robust and non-robust statistics is important when analyzing small sample data because in small samples, a few very high or low scores might substantially shift the mean, and hence one's conclusions about the location of the variable.

1.2.2 Variability Statistics

Variability (also called spread or dispersion) **statistics** capture how much values vary or spread out along the scale of measurement. Some statistics measure variability "around," or in reference to, a location statistic, whereas others simply render variability as a value. Let's look at four statistics for learning about the variability of a numeric variable.

- The **variance** finds how far, on average, values deviate from the mean. But because the variance squares all the deviations first, the statistic is in squared measurement units and can be very difficult to interpret. The concept of the sample variance is made clear by the operations in Formula 1.2: squared deviations from the mean are summed and then averaged (with $n-1$ rather than n). Why do we sum all n of the squared deviation scores but only divide by $n-1$? Because when we find the mean of a group of n scores, all n scores are *free to vary* (meaning they could take on any value). But when we find the mean of a group of scores that are themselves calculated using the sample mean (such as the scores in the numerator of Formulas 1.3 and 1.4), only $n-1$ scores are free to vary. For an example, imagine a sample of four scores with a mean of 4.0. We arbitrarily say $x_1 = 5$, $x_2 = 2$, and $x_3 = 6$. If the sample mean is known, which it is, x_4 is not free to vary; it must be 3 if the mean is to equal 4. When we calculate the deviation scores in the variance formula, the sample mean is known and used in the calculation (see Formula 1.3). Therefore, we must divide by $n-1$ to get an accurate average of those scores.
- The statistical properties of the variance are similar to the mean. That is, the variance is a sufficient statistic because it incorporates all n scores but is non-resistant because it can be influenced by extreme values, particularly when n is small. In fact, relative to the mean, the variance is actually a less resistant statistic because the influence of any extreme score is *squared* in the variance calculation.
- The **standard deviation** (SD) is simply the square root of the variance (see Formula 1.4). The SD is a more interpretable variability statistic. By taking the square root of the variance, average variability around the mean is returned to the variable's original measurement units. The SD inherits the variance's sufficiency and non-resistance.

$$s^2 = \frac{\Sigma(x-\bar{x})^2}{n-1} \tag{1.3}$$

$$s = \sqrt{\frac{\Sigma(x-\bar{x})^2}{n-1}} \tag{1.4}$$

- The **median absolute deviation** (MAD) finds the deviation of each score from the median, renders those into positive numbers with an absolute value operation, and then finds the median of those scores (see Formula 1.5). The MAD therefore finds the middle deviation score, rather than the average deviation score represented by the SD. The MAD is sufficient (because it incorporates all values) and more resistant than the SD for the same reasons that the median is more resistant than the mean. It is worth noting that in perfectly normal data (think: a "bell curve"-shaped distribution of values), the SD and MAD are identical statistics and will provide the same value. The presence of extreme values in either or both ends of the distribution, however, affects the SD more than the MAD, which is why comparing those statistics helps us learn about the presence of extreme scores in the data.

$$MAD = median\left(\left|x_i - med\left(x\right)\right|\right)$$ (1.5)

- The **interquartile range** (IQR) is the difference between the 3rd quartile (Q_3, or the 75th percentile) value and the 1st quartile (Q_1, or the 25th percentile) value (see Formula 1.6). The IQR describes the range of scores for the middle 50% of the values.

$$IQR = Q_3 - Q_1$$ (1.6)

1.3   Data Analytic Example 1

To demonstrate the R functions for generating the location and variability statistics in the previous section, we will work with data from the 2011–2012 National Health and Nutrition Examination Survey (NHANES), which is in the NHANES R package. In the example below, we generate statistics for summarizing the weight variable which is measured in kilograms. All of the code in the example below is in ASRRCh1.R.

The chunk of output below shows:

1. Library the NHANES package and then create a data object (called *dat*) by retrieving just the data from the 2011–2012 survey.
2. The location and variability statistics for weight with the R functions that produce them. Notice that some functions in R will not work if there are missing data (e.g., empty cells) in the data frame. When that happens, we add an na. rm. = TRUE (or T) argument to the function.
3. There is no built-in R function for finding the mode of a numeric variable. We could write a function to find the mode, but it would still deliver only the single most frequent value. Below we generate a table of the most frequent values which will show us the mode and allow us to explore possible multimodality.

```
#Library needed packages
library(NHANES)

#create small dataset by selecting survey year and sampling from those cases
dat<-NHANES[NHANES$SurveyYr=="2011_12",]

#generate location statistics
#na.rm=T tells the function to ignore missing data
mean(dat$Weight,na.rm=T)

## [1] 70.32769

median(dat$Weight,na.rm=T)

## [1] 72.1

mean(dat$Weight,tr=.1,na.rm=T)

## [1] 71.01908

#explore multi-modality
#create a frequency table,  sort it by frequency, then print just the first 6
rows of the table (i.e., the most frequent values)
library(dplyr)

tab<-table(dat$Weight)
sorted_tab <- tab %>%
  as.data.frame() %>%
  arrange(desc(Freq))
head(sorted_tab)

##   Var1 Freq
## 1 73.8   22
## 2 65.3   20
## 3 61.7   19
## 4 84.8   19
## 5 61.5   18
## 6 78.1   18
```

What do we learn about the typical weight of participants in the NHANES study from these location statistics? The average, or mean, weight is 70.3 kgs (which is about 155 lbs., by the way). The more resistant location statistics (median = 72.1 kg, 10% trimmed mean = 71.0 kg) are a little higher. This indicates that either that the only extreme values are below the mean, or that the "pull" of the extreme values below the mean is a little stronger than those above the mean. Remember, resistant statistics are designed to resist the influence of extreme scores on the summary statistic, regardless of where they are. The modal value, or the most common weight, is 73.8 kg. Our sorted frequency table indicates that there is a prominent secondary mode of 65.3 kg, which suggests that the distribution of weight may have a distinct subgroup with a much lighter typical weight.

Now let's generate the variability statistics covered earlier, keeping in mind that the variance and standard deviation are equivalent statistics and the SD is easier to interpret, we only generate the SD. Each statistic has its own function in R.

```
sd(dat$Weight,na.rm=T)

## [1] 28.80093

mad(dat$Weight,na.rm=T)

## [1] 24.16638

IQR(dat$Weight,na.rm=T)

## [1] 32.7
```

What do we learn about the variability of weight values from the statistics above? The SD indicates average variability in weight values is about 28.8 kg from, or around, the mean. The MAD is smaller (24.2 kg), which is likely due to the fact that the MAD is less influenced by very small or large values. Finally, the range of scores for the middle 50% of the students (IQR) is 32.7 kg. In other words, 32.7 kg separate the 1st and 3rd quartile (or 25th and 75th percentile) values.

With our statistical summary done, we now turn to graphical summary methods. Plots are a necessary part of any descriptive data analysis, and an excellent complement to statistical summaries for two reasons. First, although they cannot offer the precision of statistics, they are superior to statistics for providing context for a summary statement such as a location or variability statistic. Second, plots are valuable tools for exploring data because they allow us to see our data through graphical lenses, with less (or sometimes no) numerical summarizing. Plots allow us to see our data in a more granular way, including its complexities and patterns, whereas statistics are designed to summarize over that complexity. Three graphical tools are useful for summarizing and exploring data from a numeric variable: the **histogram**, **density plot**, and **boxplot**. Let's use them to display and summarize the weight variable in the 2011-12 NHANES survey data and discuss their features and value to descriptive analysis.

A **histogram** groups values into equal-interval bins along the x axis and represents the frequency of that case in that bin of values on the y axis. The hist() function in R sets the bin interval by following two rules: achieve about ten bins and maintain "pretty" break points. We can retrieve those values (the break points and the number of bins) by assigning the histogram output to an object and calling that object (see output below). Notice that we suppressed the plot output, as we already produced that.

```
hist(dat$Weight)
```

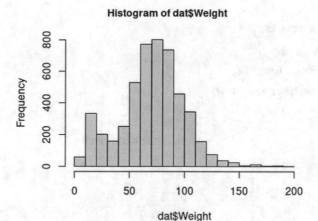

```
h=hist(dat$Weight,plot=F)
h$counts
```

```
## [1]  59 334 202 159 252 529 773 802 737 457 344 157  75  37  25   5  12   4   5
## [20]   1
```

```
h$breaks
```

```
## [1]   0  10  20  30  40  50  60  70  80  90 100 110 120 130 140 150 160 170 180
## [20] 190 200
```

This output shows that by default R created a histogram with 5-point intervals, displayed in the "breaks" output. Maintaining these convenient intervals required a histogram with 12 bins; the "counts" output shows the number of cases in each bin. The histogram becomes a better tool for exploring a numeric variable when we can control the number of bins ourselves and adjust it to do more or less summarizing of the data. Here are two examples of histograms of the weight data, the first with a wide bin interval (in which we suggest to R a goal of 5 bins) and the second with a narrow interval (in which we suggest to R a goal of 50 bins). Remember that R tries to reconcile our "breaks=" instruction with its other mandate to generate break points with round numbers.

```
h1<-hist(dat$Weight, breaks=5)
```

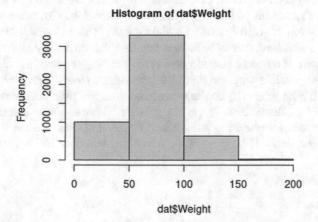

```
h1$breaks
## [1]   0  50 100 150 200
h1$counts
## [1] 1006 3298  638   27
h2<-hist(dat$Weight, breaks=50)
```

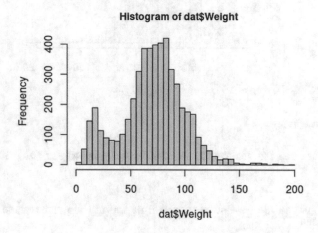

```
h2$breaks
##  [1]   0   5  10  15  20  25  30  35  40  45  50  55  60  65  70  75  80  85  90
## [20]  95 100 105 110 115 120 125 130 135 140 145 150 155 160 165 170 175 180 185
## [39] 190 195 200

h2$counts
##  [1]   7  52 145 189 113 89 81 78 101 151 219 310 386 387 398 404 420 317 268
## [20] 189 176 168  92  65 47 28 17  20  19   6   3   2   6   6   4   0   3   2
## [39]   0   1
```

Let's evaluate these plots as summary tools. The first histogram (four bins with 50 kg intervals) does too much summarizing, by which we mean that it combines a large range of scores into a single bin. An over-summarized plot doesn't allow us to learn much about the distribution of weight scores. This is because the scores are restricted to a small number of wide bins and thus the variability of scores *within* a bin is unknown. If our data analytic goal is to "see" the shape of the distribution in the smoothest possible way, the first histogram is not ideal. The second histogram (42 bins with 5 kg intervals) on the other hand does too little summarizing, giving us a picture of the distribution that is too granular. These examples demonstrate that the bin interval is a **smoothing parameter**. Collapsing data into too few bins produces an *oversmoothed* histogram. Oversmoothing can obscure, or even eliminate, characteristics of the variable's distribution that are important to see and explore. Too many bins, on the other hand, preserve too much of the individual scores and produce an *undersmoothed* histogram. Undersmoothing allows the plot to display random idiosyncrasies (sometimes called noise) in the sample data that can, again, obscure the true shape of the variable's distribution. The best histogram finds the optimal amount of smoothing to "see" the distribution's true shape.

With some exploratory adjusting of the smoothing parameter (using the *breaks* argument in the hist() function), we might arrive at an optimal bin interval of 5 kg. This histogram seems to give us the best picture of the distribution of weight values; it doesn't oversmooth or undersmooth (see plot below). Once we have decided on the optimal bin interval value, then we should complete our plot with labels and titles that make it more reader-friendly and appropriate for presentation. Notice that the code to add axis labels and titles to a plot merely requires some additional arguments within the hist() function.

```
h<-hist(dat$Weight, breaks=20,
        prob=T,
        main = "Weight of NHANES participant",
        xlab = "weight (kg)")
```

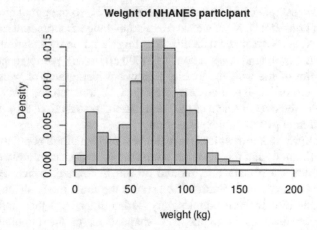

What do we learn about the distribution of the weight of NHANES survey participants from this plot? Most of the weight values are between 50 and 100 kg, and indeed that range includes all of our location statistics. However, there is considerable variability in weight values around those typical values, and evidence of a separate "bump" of values between 5 and 15 kgs.

A **density plot** is a natural extension of the smoothing function of a histogram to display the true shape of the distribution of values. To understand a density plot, we first must think about our data almost always being a sample or subset from a larger

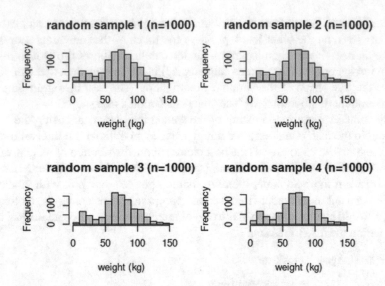

Fig. 1.1 Histograms of travel time to work for four random samples of 2000 counties

population. In the examples above using the 2011-12 NHANES survey, the data was from a random sample of 5,000 American households, so the population would be all American households. Next, we have to acknowledge that random samples from a population vary. We can simulate this sampling variation by sampling 1,000 participants from our pretend "population" of 5,000 NHANES participants and generate a histogram of the weight variable. Below are histograms of weight for four random samples of 1,000 participants (see Fig. 1.1). You can see that although the samples reflect the overall shape of the "population" distribution, they each present a slightly different picture of the population.

So, how are we to know which one is the best picture of the population distribution? We don't know, and this is the problem that a density curve addresses. A density curve renders a smooth line, plotted on our histogram, which *estimates* the shape of the population distribution (i.e., shows the most likely shape). A density plot is valuable tool because samples vary—sometimes by a lot—and any single sample may mislead us as to the true shape (i.e., in the population) of the distribution.

How does density curve estimation work? Let's walk through it, using a simple dataset of 6 points which are shown in the left panel of Fig. 1.2 below. The density curve for this sample data (blue curve in the plot on the right) is estimated with the help of a weighting function (see the red curves in the plot on the right). The weighting function is simply a normal curve whose mean is placed over a *focal* x value. Then, any value that falls within the "neighborhood" of that curve is assigned a weight. The focal value gets full weight and adjacent values get less weight,

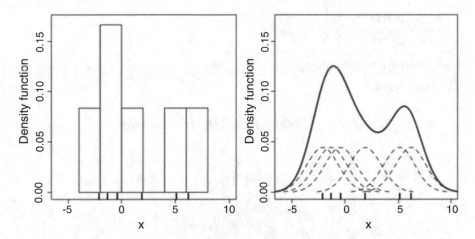

Fig. 1.2 Histogram (left panel) and density plot with normal weighting curves (right panel)

proportionate to how far they are from the focal value. Then, the weights are summed to arrive at an estimated density (or probability) for each focal value. When those estimated densities are connected with a line, you end up with a density curve. This procedure utilizes the data around a particular focal value, weighted accordingly, to help inform and estimate the probability (or density) of that value in the population (Fig. 1.2).

Density estimation, then, follows a pretty intuitive idea. If for example you wanted to predict the weather for June 1 of next year from this year's weather on that date, you would do well to consider weather in the days surrounding June 1 to develop your prediction, with the days just before and after June 1 having the most relevant (or most strongly weighted, in density estimation terms) information. The *particular* weather on June 1 will vary (randomly, of course) year over year, but the *general* weather in the early days of June is much less likely to vary year over year and therefore is a much better estimate of next year's weather.

The plot below is our histogram of weight from the NHANES participants with a density curve added. You can see that the density curve smooths out the blocky function of the histogram. It estimates the likely shape of the distribution of weight in the population of American households—assuming that our sample is a random sample from the population.

```
X<-na.omit(dat$Weight)
h<-hist(X, breaks=20, na.omit=T,
        prob=T,
        main = "Weight of NHANES participant",
        xlab = "weight (kg)")

lines(density(X))
```

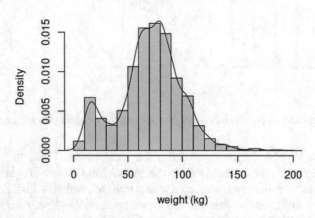

The **boxplot** is another valuable tool for visualizing the distribution of a numeric variable. A boxplot (also known as a box-and-whisker plot) is built around a 5-number summary of a numeric variable. The five numbers consist of the lower adjacent value, Q_1, Q_2, Q_3, and the upper adjacent value (see boxplot below). Adjacent values are located by subtracting the quantity 1.5 x IQR from the 1st quartile value and adding it to the 3rd quartile value. If the span from the quartile to the adjacent value includes the minimum or maximum value, the adjacent value simply takes that value. Let's generate a boxplot of our NHANES sample weight data.

```
b<-boxplot(dat$Weight,horizontal=T)
```

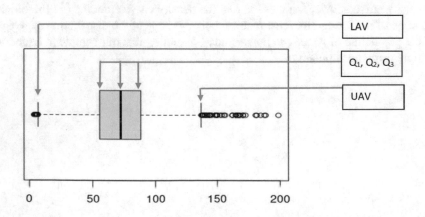

```
#5-number summary: LAV,Q1,Q2,Q3,UAV

b$stats

##         [,1]
## [1,]    6.7
## [2,]   55.4
## [3,]   72.1
## [4,]   88.1
## [5,]  136.9
```

```
#generate boxplot rule-defined outliers, list in ascending order, and count the
number of scores
sort(b$out)

##  [1]    3.6    4.2    4.5    4.7    4.8    4.8    4.8    5.1    5.3    5.3    5.4    5.5
## [13]    5.6    5.6    5.8    5.9    6.2    6.2    6.2    6.2    6.2    6.3    6.3  137.3
## [25]  137.3  138.0  138.2  138.4  138.4  138.7  138.7  138.7  138.7  138.7  140.1  140.1
## [37]  140.1  140.1  140.1  140.1  140.1  140.1  141.4  141.4  141.4  142.3  143.6  143.6
## [49]  143.6  143.6  143.6  143.6  145.0  145.5  148.2  148.2  149.1  149.8  149.8  150.1
## [61]  152.3  153.7  155.4  156.4  162.0  162.2  162.2  164.9  164.9  164.9  166.5  166.5
## [73]  166.5  167.6  167.6  168.1  170.5  170.5  170.5  172.5  180.6  181.4  184.5  187.5
## [85]  188.5  198.7
```

```
length(b$out)

## [1] 86
```

As shown in the example above, we can retrieve the 5-number summary from the boxplot() function output by saving the output in an object and then retrieving it with the "stats" element, as shown above. The lower adjacent value (LAV), 1st quartile value ($Q1$), median ($Q2$), 3rd quartile value ($Q3$), and upper adjacent value (UAV) are listed in lines 1–5 of the output. As a tool for exploring the distribution of a numeric variable, the boxplot has limitations and strengths. A boxplot does not display the mean and standard deviation, which are common location and variability statistics. On the other hand, a boxplot is built on resistant location (median) and variability (IQR) statistics. This quality makes the boxplot an excellent tool for identifying potentially influential values, which is our final descriptive analysis task.

Identifying extreme values is an important data analytic task for two reasons. One, extreme values have the potential to disturb non-resistant summary statistics such as the mean and SD, so we would like to anticipate and deal with those disturbances. Two, extreme values (as well as multimodality in a distribution) suggest that our sample data may contain elements from other populations than the assumed population of interest. Keep in mind that extreme values might be incorrect values, and that possibility should be eliminated first. Assuming however the values are correct in the example above, the people with extremely low and high weight shown in the boxplot above might be different in some way(s) from the typical American household participants in the NHANES survey.

An **outlier** is a value that exerts disproportionate influence on summary statistics. Diagnosing an extreme value as an outlier must be done with an objective and replicable process. It should also be done with a process that is robust; in other words, the diagnostic method for identifying outliers should not be influenced by those same values. The **boxplot rule** meets both diagnostic criteria. The boxplot rule defines an outlier as any value that is either lower than the lower adjacent value or higher than the upper adjacent value. Because adjacent values are based on the median and IQR—both resistant statistics—the decision rule itself is not influenced by extreme values. Applying the boxplot rule to our weight data shows 86 outliers, 23 of which are low outliers, meaning they are less than the lower adjacent value (6.7 kg). Most of the outliers are high outliers and have values greater than the upper adjacent value (136.9 kg).

1.4 Data Analysis with a Categorical Variable

In the example above, we used statistical and graphical tools to summarize a numeric variable. But many interesting and important variables are not numeric, such as a diagnostic test result (positive or negative) or a college application decision (accepted or not accepted). In the social sciences, opinion and attitude measurements often result in categorical data (e.g., "Have you ever smoked marijuana?" yes/no). These are examples of categorical variables, and data from categorical variables aren't "scores" but frequencies, where each person or case in the study is counted in one of the variable's categories.

With numeric variables, families of summary statistics are based on a location statistic. For example, the mean is the basis for the standard deviation and the median forms the basis of the IQR. With frequency data, the statistic from which many other statistics are built is the proportion. A **proportion** is simply the number of cases in a category as a fraction of the total cases, in decimal form. Proportions and probabilities are very close concepts. Consider this example: we get a random sample of ten individuals and measure their dominant hand (right-handed or left-handed). Imagine that we are interested in summarizing or exploring left-handedness. So left-handedness will be our "presence" category (coded 1) and right-handedness becomes the "absence" category (coded 0). The hypothetical sample data are 0,0,1,0,0,0,1,0,0,0. If you find the mean of these "scores" (2/10 = 0.20), you have found the proportion left-handers in the sample. The 0.20 also is the probability that a randomly selected person from the sample will be left-handed. With a two-category variable, summarizing the "presence" category also summarizes the "absence" category, or the proportion of right-handers (0.80). Notice how the correct interpretation of the probability depends on the category of interest, or "presence," category. Defining the "presence" category will become important when we do bivariate analysis with a categorical outcome variable in Chap. 4.

In studies with categorical variables, the "distribution" we are interested in describing and exploring is the distribution of frequencies or proportions across the levels, or categories, of the variable. The statistical tools that allow us to summarize a distribution of frequency data are tabular: tables of frequencies by category and proportions (or percentages) by category. The graphical equivalent of a frequency or proportions table is the barplot. Barplots use bars to display frequency or proportion data but maintain space between the bars on to acknowledge that variable's categories are discrete and nonoverlapping. Furthermore, there is no quantity implied in the ordering of bars in a barplot. These summary tools are displayed in the example below.

1.5   Data Analytic Example 2

To demonstrate the statistical and graphical tools for summarizing and exploring a categorical variable, we will work with the Race1 variable in the 2011–2012 NHANES data. All of the code in the example below is in ASSRCh1.R.

The chunk of output below shows:

1. Library the janitor package. This allows us to use the tabyl() function for producing our table of frequencies and proportions.
2. Create a table of percentages by category for plotting and use the barplot() function to generate the plot.

```
#frequency table with proportions
library(janitor)


tabyl(dat$Race1)

## dat$Race1      n  percent
##       Black  589   0.1178
##    Hispanic  350   0.0700
##     Mexican  480   0.0960
##       White 3135   0.6270
##       Other  446   0.0892

#create table of category percentages for plotting, then plot
tab<-prop.table(table(dat$Race1))*100
barplot(tab,ylim=c(0,100),ylab = "Percentage(%)",las=2)
```

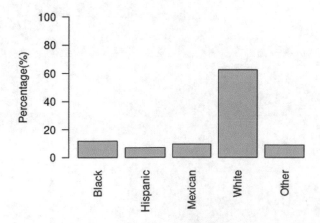

What do we learn about the distribution of race from this table and plot? First, about 63% of the survey participants are White. Blacks constitute about 12% of the sample, and people identifying as Hispanic, Mexican, or other make up the remaining 25% of the sample. Finally, "seeing" the distribution of a categorical variable through a barplot can help us decide if and how to recode a categorical variable (e.g., combine or drop categories) into a new variable for analysis. This skill will show up in Chaps. 3 and 4. Before we get there, however, we must learn how the design of a study produces data figures into what statistics we use and how we interpret those statistics (Chap. 2).

1.6 Problems

The problems below all use datasets in the MASS package. Use library(MASS) *to make the datasets available and* ? datasetname *to see the documentation for each dataset, which will provide variable names and measurement details.*

1. In the cats dataset, generate the following location statistics for the Bwt variable: mean, 10% trimmed mean, and median. Do the resistant and non-resistant statistics differ substantially? What does that suggest about your data?
2. In the cats dataset, find the mode of the Bwt variable. Generate a frequency table and determine if there are multiple modal values (other Bwt values that have nearly as many cases). Are those modal values close together, or separated by other values? If modes are separated by other values, what might that suggest about your data?
3. In the birthwt dataset, find the standard deviation of mother's age. Interpret it. Generate a boxplot of the age variable. Using the boxplot rule, identify any extreme values (outliers). Explain how deleting that value would affect the standard deviation. Then, use sd(birthwt$age[birthwt$age < 45]) to exclude that value in the SD calculation. How much did the SD change? Repeat this analysis with the IQR, and explain why the IQR is not affected by removal of the outlier.
4. In the Melanoma dataset, generate a histogram of survival time (let R decide on the bin interval). Use breaks = 5 and breaks = 20 to generate two more histograms of the survival time data. Which one oversmooths? Undersmooths? In terms of insights the plot provides into properties of survival time, what is lost between the first (default bin width) and the breaks = 5 histogram? Compare the default histogram to the breaks = 20 histogram: What are you seeing in the breaks = 20 plot that may detract from your learning about survival time?
5. Generate a probability histogram of survival time from the Melanoma dataset (use the default bin interval) and add a rugplot and a density plot to the histogram. What does the density plot show you about the distribution of survival times in the population of melanoma patients?
6. Generate a boxplot of survival time from the Melanoma dataset. Find and interpret the 5-number summary from the boxplot output. If there are boxplot-defined outliers in the data, find their values too. From the boxplot, does it appear that survival time data are skewed, and if so in what direction? Check that with a skew statistic.
7. In the birthwt dataset, identify any outliers in mother's age using the boxplot rule.
8. Using the UScereal dataset, do a descriptive analysis of the calories variable. Generate location and variability statistics as well as a plot of the variable's distribution. Summarize your findings in a few sentences.
9. In the cats dataset, generate frequency and proportion tables to summarize cat sex. What proportion of cats are female?

Chapter 2
Foundations II: Statistical Models and Study Design

Covered in This Chapter
- Statistical models and their importance to data analysis
- Measurement scales and variables
- Study design elements and their importance to data analysis

2.1 Goals of Data Analysis with Bivariate Data

In Chap. 1—our first foundations chapter—we introduced some basic statistical and graphical tools to summarize and explore data from a single variable and learned how to use R to conduct that analysis. A working understanding of basic statistics and statistical computing is foundational for doing data analysis with more complex data. In this chapter—our second foundation chapter—we look closely at the study that generated the data we wish to analyze. It turns out that we can't properly analyze data, nor can we accurately interpret data analytic results, without understanding the design, measurement, and statistical model of the study from which the data came.

In Chap. 1 our data analysis focused on data from a single variable, or univariate analysis. We learned what properties of a numeric variable are important to summarize and explore and how those tasks translate to a categorical variable. Research questions, however, almost always consist of two (or more) variables. Here are some examples:

- Are extracurricular activities in high school related to college acceptance?
- Do introverts or extraverts have more social media friends and contacts?
- How is daily stress related to sleep quality?

With research questions like these, data analysis focuses on summarizing and exploring *relationships* between variables. The goals of data analysis in bivariate

B. Blaine, *Introductory Applied Statistics*, https://doi.org/10.1007/978-3-031-27741-2_2

data are the same as in univariate data: to summarize and explore. Let's look at each goal in a bit more detail.

2.1.1 Summarizing Bivariate Data: Relationship Direction and Magnitude

Summarizing a relationship between variables requires that we establish the *direction* of the relationship. To say that two variables are related, or covary, is to say that changes in one variable are systematically related to changes in the other variable. But do values from one variable generally *increase*, or *decrease*, with changes in values from the other variable? Here's an example: say we measured daily stress and sleep quality in a sample of students. Let's stipulate that the two variables are related—that changes in stress are associated with changes in sleep quality. If we found that people who reported higher stress also had better sleep quality, then there would be a positive relationship in the data (i.e., sleep quality increases with increasing stress). If however increased stress levels were associated with poorer sleep quality, the relationship would be negative (i.e., sleep quality decreases with increasing stress).

In addition to the direction of the relationship, we must also establish the *magnitude*, or strength, of the relationship. Knowing the direction doesn't tell you *how much* change in sleep quality is associated with a unit increase in stress. And it's the "how much?" question that is important for evaluating the practical significance of a relationship. For example, if high, compared to low, daily stress was associated with a 0.1 hour difference in sleep quality (or 6 minutes a night), we might not be as concerned with reducing daily stress as we would be if the relationship magnitude was 1 hour per night.

There are many statistics whose purpose is to summarize the direction and magnitude of a relationship between variables. Some statistics are designed for, or are much more interpretable in, a particular statistical model. We introduce statistical models below and see how essential they are to data analysis and interpretation. Before we do that, let's address the second data analytic goal—exploration.

2.1.2 Exploring Bivariate Data

Exploring bivariate data is important for the same reasons we explored data from a single variable in Chap. 1. Statistical summaries are necessary but not sufficient to capture the nature of a relationship between variables. Statistics reduce the complexity of data, which is helpful, but can overlook aspects of the relationship that go beyond, or even contradict, statistical summaries. Let's look at two examples of how exploration can help us do better data analysis and come to more accurate conclusions about bivariate relationships.

Example 1 In this example we use the 2011–2012 NHANES data and explore the relationship between gender and the number of days in a typical week that participant does moderate or vigorous-intensity activity. The plotting was done with the lattice package, which has some plotting functions that we will use in later chapters too.

```
library(NHANES)
dat<-NHANES[NHANES$SurveyYr=="2011_12",]
library(lattice)
bwplot(Gender~PhysActiveDays,data=dat)
```

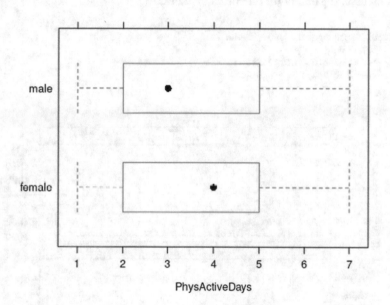

```
tapply(dat$PhysActiveDays,dat$Gender,mean,na.rm=TRUE)
```

```
##   female       male
## 3.878965 3.759259
```

```
tapply(dat$PhysActiveDays,dat$Gender,median,na.rm=TRUE)
```

```
## female    male
##      4       3
```

Let's consider the typical or average number of days per week of physical activity, separately for females and males. The boxplots show similar distributions in terms of the middle 50% (IQR), but the medians are very different. When we generate both resistant and non-resistant location statistics, we see that the mean level of physical activity is about the same for females and males (they differ by about 1/10th of a day), but the medians differ by a full day. This suggests that *within* the

IQR (which we can't see from a boxplot), physical activity data is distributed differently for females than males. How would we characterize the relationship between gender and physical activity based on this analysis? Inasmuch as the median, compared to the mean, difference is the more robust location statistic, that difference (or 1 day per week) is probably is the more accurate indicator of the true relationship, with females being more physically active than males.

Example 2 In this example we use the 2011–2012 NHANES data and explore the relationship between gender and age at which the participant first started to smoke cigarettes fairly regularly. Once again we will use lattice functions for the plotting. In this example let's consider the typical or average variability in the age of first smoking data, separately for females and males.

```
histogram(~SmokeAge|Gender,data=dat,
        type="density",
        breaks=seq(0,80,by=2.5),
        layout=c(1,2))
```

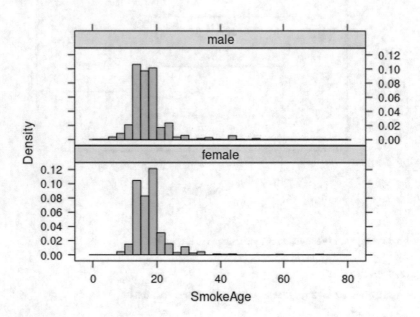

```
tapply(dat$SmokeAge,dat$Gender,sd,na.rm=TRUE)

##    female      male
## 5.433583 5.767121

tapply(dat$SmokeAge,dat$Gender,mad,na.rm=TRUE)

##   female     male
## 2.9652 2.9652
```

The histograms above show that, although most people start smoking in their teens and 20 s, there appear to be more men than women who start smoking as older adults (see the longer positive tail of the distribution in the histogram for males). This suggests that average variability in age of first smoking would be greater for males than females. When we generate variability statistics, we see a higher SD in males than females (by 0.3 years from the mean), but the MADs are the same. So although the histograms suggest more variability in males' age at first smoking, due mainly to a small number of high values, a resistant variability statistic reveals that the variability is about the same in the two groups.

Echoing the principles we learned in Chap. 1, these examples illustrate that statistical summaries of relationships can be misleading, particularly when they are captured with only non-resistant statistics. Exploratory data analysis allows us to "see" bivariate relationships through different statistical lenses and learn which statistics might describe the true relationship more accurately.

2.2   Statistical Models and Data Analysis

The data analytic examples show how summarizing a relationship between variables can require different statistics, different plots, and even a different vocabulary for describing the relationship. Statisticians and analysts are commonly presented with data and a research question and expected to address or "answer" that question with the data. But how do we know what statistics, plots, and other procedures to use in data analysis? The answer is in the statistical model underlying the research question and variables. A **statistical model** is a mathematical statement of a relationship between variables. It connects the variables in our research question, on the X and Y sides of a linear equation, and incorporates the measurement scales of (i.e., the type of data we have in) those variables. The four statistical models we will learn and use in this book are all subtypes of a general model called, unsurprisingly, the **general linear model**. The general linear model takes the form

$$Y = \beta_0 + \beta_1 X + \varepsilon$$

where

- Y is the **outcome** or dependent variable
- X is the **predictor** or independent variable
- β_0 is the intercept or the value of Y when $X = 0$
- β_1 is the function that represents the best linear relationship between X and Y
- ε is the error associated with predicting (or explaining) Y from X

In any research scenario, we have X and Y data courtesy of the survey or study that collected it. Our focus in data analysis is on the parameters β_0, β_1, and ε. A **parameter** is a number that describes a population, whereas statistics are numbers that describe samples. Because we analyze sample data, which is very likely only a small part of some larger population, the values of those parameters are unknown. However, we can *estimate* them from our sample data, on the logic that if the sample reflects the population, the sample-based statistic should be a reasonably accurate estimate of the unknown population parameter.

Of the three parameters in the equation above, β_1 is our primary focus in data analysis because it captures the relationship between X and Y. In other words, β_1 (or, more accurately, our sample-based *estimate* of β_1) describes the direction and magnitude of the relationship between X and Y. So β_1 answers the question "how, and how much, does Y change with a unit change in X?" And that's what most research questions are interested in. The intercept (β_0) becomes important only when we want to use the linear equation to predict someone's score on Y from their particular X value. Any errors in the model are assumed to be random, with positive and negative errors balancing each other out across participants, so ε is assumed to be zero.

Let's move away from mathematical terms and talk about the general linear model in conceptual terms. Conceptually, the equation can be thought of as

$$\textbf{Data} = \textbf{Model} + \textbf{Error}$$

where

- **Data** represents our outcome variable—the variable we want to explain or predict—as observed through sample data. Keep in mind that sample data may give us less-than-perfect observations of the outcome variable as it exists in the population.
- **Model** represents our "construction" or understanding of the outcome variable, in terms of our explanatory or predictor variable. Models are always hypothetical, tentative, and reliant on the data to validate or invalidate them.
- **Error** represents any lack of fit between model and data. Lack of model fit can arise from a variety of sources, but there are two common ones we can keep in mind as we analyze data in the coming chapters: poor measurement of the outcome and/or predictor variable and an incomplete model of the outcome variable.

The combination of X and Y variables *and their respective measurement scales* yields a particular form of the general linear model, and it is that model that directs our data analysis toward the proper statistics, plots, and interpretations. We dig into those models below, but first we must review measurement scales—and the type of data each scale produces—so that we can correctly diagnose the statistical model inherent in our data. As summarized in Table 2.1, the four measurement scales are hierarchically ordered, with each having the dimension(s) of the scale(s) beneath it. **Nominal** scales simply name or categorize values; they have no quantity associated with them. Even though a nominal variable might have a numeric code (e.g., 0 = female, 1 = male), the numbers are simply names—they imply no quantity or ranking of the levels of gender. **Ordinal** scales categorize and rank values. Ordinal

Table 2.1 Measurement scales

Measurement scale	Type of data	Measurement function and examples
Nominal	Qualitative	A scale that names or categorizes cases only; no quantity is assumed, e.g., gender, political affiliation
Ordinal	Quantitative	A numeric scale that ranks cases only; equal intervals are not assumed, e.g., class rank, 5-star rating scale
Interval	Quantitative	A numeric scale where numbers indicate the amount of the measured variable; equal intervals are assumed; no true zero, e.g., standardized test score, Likert rating scale
Ratio	Quantitative	A numeric scale where numbers indicate the amount of the measured variable; equal intervals and a true zero are assumed, e.g., height, income

variables are numeric, but the differences between the ordinal values—called intervals—cannot be assumed to be equal. For example, if class rank measures academic
achievement, the difference between the class valedictorian (rank 1) and salutato-
rian (rank 2) may be any value, as could the interval between any other two students
with adjacent ranks. Ordinal scales, as a result, add ranking to naming.

Interval scales, in addition to naming and ranking, are quantitative scales that
maintain equal intervals. In other words, the "amount" of the variable being measured is assumed to be the same between adjacent values anywhere on the scale.
Interval scales might or might not have a zero on them, but the zero is not a *true* zero
in the sense that it does not represent absolutely none of the variable. For example,
on a 0 (strongly disagree) to 4 (strongly agree) response scale, the zero does not
mean zero agreement. Finally, **ratio** scales are quantitative scales that name, rank,
maintain equal intervals, *and* have a true zero. With ratio variables such as height
and weight, zero actually indicates no height or weight.

Let's simplify this situation a bit so that we end up with a workable set of statistical models that apply to most research scenarios involving two variables (predictor
and outcome). First, ordinal variables are uncommon (or are not of interest to
researchers, or both), so relatively few statistical procedures have been developed to
analyze ordinal data. We sacrifice very little in excluding ordinal X or Y data from
our set of statistical models. Second, statistical procedures that require quantitative
data don't distinguish between interval and ratio scale data. In other words, the statistical procedure will not "know" if the zero in the data is a true zero or not. The
statistics we use to summarize quantitative numeric data (which, by the way, can be
integer or decimal values) require equal intervals, and both interval and ratio scale
data assume that quality. So, although we can treat interval and ratio data as the
same for data analytic purposes, we must know the difference for interpretation
purposes. Once we apply these practical realities, we end up with two data types
that cover most research variables: categorical (nominal scale data) and numeric
(interval or ratio scale data).

With that foundation, we now turn to the statistical models covered in this book.
Modifying the general linear model with each combination of the data types

Table 2.2 Common forms of the general linear model

General linear model form	Traditional alternate names	Chapter
numeric $Y = \beta_0 + \beta_1 categorical\ X + \varepsilon$	ANOVA model	3
categorical $Y = \beta_0 + \beta_1 categorical\ X + \varepsilon$	Proportions model	4
numeric $Y = \beta_0 + \beta_1 numeric\ X + \varepsilon$	Regression model	5
categorical $Y = \beta_0 + \beta_1 numeric\ X + \varepsilon$	Logistic model	6

rendered above produced a particular statistical model (see Table 2.2). Each of these models is very commonly used in research in many disciplines. The models are associated with traditional names, but the form of the statistical model (i.e., what type of data are in the X and Y variables) is much more important than its name, because the model form guides the statistics, plots, and other statistical methods used to estimate β_1 in each model.

The general linear model is a powerful tool for organizing statistical models by their constituent variables and data types and for providing guardrails for our data analytic work. It can be extended to accommodate much more complex models than those we are learning in this course, but these four basic models are the foundation for all those other models. Let's get a closer look at each model.

2.2.1   ANOVA Model

The **ANOVA model** is probably the most common statistical model used in research. "ANOVA" is a shortened reference to *analysis of variance* because, traditionally, that was the default data analytic method in designs with categorical X and numeric Y. Analysis of variance is a statistical procedure that (somewhat counterintuitively) compares means from different groups. The mean difference is the most common, but not the only, statistic for summarizing an X-Y relationship in this model. What kind of research question is suitable for an ANOVA model?—one in which groups of participants are compared on some dimension of a numeric variable. Here are two examples:

- What is the effect of a new antianxiety drug treatment, compared to a placebo, on a measure of anxiety symptoms (as measured by a symptom checklist with 30 symptoms)?
- Do left-hand dominant compared to right-hand dominant people differ in their texting speed (as measured by characters per minute)?

As these examples show, research questions in ANOVA model studies rarely specify the statistical dimension of Y predicted to be associated with X. General phrases like "is there an effect of treatment?" or "do groups differ?" do not state or imply a statistical lens to summarize the relationship. Often the default data analytic approach in ANOVA model studies is to look at the mean difference, or how groups differ "on average," as the estimate of β_1. But we know that the mean, and therefore the mean difference, is a non-resistant statistic and vulnerable to the influence of extreme values. Moreover, many treatments and interventions also affect the variability of Y scores. This is why the data analyst needs to be open to what the data might say about the appropriate statistical lens for summarizing a location difference in Y.

One final point: in terms of the design of ANOVA model studies, the predictor variable (X) can be composed of groups that are arranged by random assignment or of groups that are selected from existing populations. This is a hugely important design element to look for in the study documentation and, if possible, confirm. The nature of the predictor variable (randomly assigned or selected) does not affect the statistical analysis but does affect the *interpretation* of the X-Y relationship found in the data. We will look at this design issue in more detail in the next section.

2.2.2   Proportions Model

A **proportions model** is similar to an ANOVA model on the X side of the model equation. We are still comparing groups; however, the outcome variable in a proportions model is not numeric (see Table 2.3). As we learned in Chap. 1, a categorical Y variable yields frequency data. Converting category frequencies to proportions controls for the size of the sample and yields a much more useful statistic. What

Table 2.3 Variable types and statistical models

		X Predictor variable		
	Data type	Categorical (k = 2)	Numeric	Model use
Y Outcome variable	Categorical (k = 2)	**Proportions model** Compares predictor groups on a categorical outcome variable	**Logistic model** Finds association between a numeric predictor and a categorical outcome variable	Classification
	Numeric	**ANOVA model** Compares predictor groups on a numeric outcome variable	**Regression model** Finds association between a numeric predictor and a numeric outcome variable	Prediction
	Statistical lens	**Comparative statistics** differences in, or ratios of, the outcome variable associated with (or caused by) the predictor variable	**Linear function statistics** change in outcome variable that is associated with a 1-unit increase in the predictor variable	

Note: k = number of categories in the variable

kind of research question is suitable for a proportions model?—one in which groups of participants are compared on "membership" in a categorical outcome variable. Here are two examples:

- What is the effect of a new job training program, compared to no training, on the probability of being employed versus not employed in 6 months?
- Are Democratic compared to Republican voters more likely to support (as measured with a yes/no response) a masking mandate policy?

As these examples show, proportions model studies investigate the probability or likelihood of being in one category or the other on Y as it relates to X. The basic statistic in the proportions model is the proportion, and it forms the basis of many other useful statistics such as risk and odds-based statistics that we will learn in Chap. 4. As with ANOVA model studies, determining whether the predictor variable is experimentally imposed on participants or selected from groups of participants that already have the quality of interest (e.g., Republican voters) is important to the interpretation of the bivariate relationship.

Because they are identical on the X side of the model equation, ANOVA and proportions models summarize X-Y relationships with **comparative statistics** (see Table 2.3). These types of statistical "lenses" come in two forms: difference statistics (e.g., mean difference) and ratio statistics (e.g., odds ratio). Although their metrics and interpretation will differ, all comparative statistics contain information about the direction and size of the X-Y relationship and therefore estimate β_1.

2.2.3 Regression Model

A **regression model** is similar to an ANOVA model on the Y side of the model equation; both models require a numeric outcome variable. A regression model, however, also has a numeric predictor. This shift to numeric X data means that a different type of statistic is required to summarize a bivariate relationship. Comparative statistics don't work because there are many different values of X in the data, not just two. Of course, an analyst could take numeric predictor data and fashion a two-group variable out of it, by, for example, dividing scores into "low" and "high" groups and then comparing those groups on Y. People do this, but in doing so, they throw away valuable information from the X data. To preserve the richness of the numeric X data, regression models (and logistic models, as we will see below) use linear functions to model X-Y relationships. Once a line is fit to numeric X/numeric Y data—a process that we will learn in detail in Chap. 5—the slope of the line becomes a very useful statistic for summarizing the bivariate relationship. What kind of research question is suitable for a regression model?—one in which a numeric predictor variable is used to explain or predict a numeric outcome variable. Here are two examples:

- Does high school GPA predict college GPA (with each measured on 4-point decimal scales)?

- Does alcohol consumption (as measured by number of drinks in a 3-hour period) explain errors on a virtual driving test?

As these examples show, regression model studies investigate how much change occurs in Y for a unit change in X. Because we use a linear function to model the X-Y relationship, the change in Y for a unit increase in X is the same across the range of predictor values. In terms of the example above, if we found that each additional drink was associated with an average of 1.5 more errors on a virtual driving test, that estimate of β_1 would describe the relationship between drinks and driving accuracy when alcohol consumption is low (e.g., going from one drink to two) or high (e.g., going from five drinks to six).

The correlation coefficient is another type of statistic that is widely used to summarize a numeric X/numeric Y relationship. Some correlation coefficients derive from the slope of a "best-fit" linear function, as explained above, and others capture an X-Y relationship differently. We will become familiar with correlation statistics in Chap. 5. Finally, the issue of whether the predictor variable groups were randomly arranged or selected, and its importance for interpretation, does not apply to a regression model.

2.2.4 Logistic Model

A **logistic model** applies to studies with a numeric predictor variable and a categorical outcome variable. This means that the logistic model shares statistical methods associated on the X side of the model equation with the regression model. It is a regression model, and, accordingly, it uses a linear function to model the X-Y relationship. However, because a logistic model requires a categorical outcome, it shares statistical methods on the Y side of the model equation with a proportions model. Accordingly, the simple proportion is the building block statistic. As we will see in Chap. 6, fitting a linear function to the probability of being in one category of Y versus another is problematic. In logistic regression this problem is solved by converting probabilities to odds and then using the log of the odds for analysis— hence *logistic* regression. We will jump into all the details of this process later.

What kind of research question is suitable for a logistic model?—one in which a numeric predictor variable is used to explain or predict a categorical outcome variable. Here are two examples:

- Does annual household income explain whether a first-year college student will return for their second year (as measured by return/doesn't return)?
- Does mother's smoking (as measured by number of cigarettes per day) during pregnancy predict low birthweight status in newborns (where low ≤5.5 lbs.)?

As these examples show, logistic model studies investigate how much change occurs in Y for a unit change in X. But since Y is categorical, a logistic model estimates the *probability* of change rather than an *amount* of change. So, for example, a logistic model would allow us to describe the change in probability of having a low (compared to normal) birthweight baby for each additional cigarette per day in mom's smoking during pregnancy.

2.3   Interpreting Bivariate Relationships: Generalizability and Causation

Our survey of statistical models and their variable types made two important points. First, summarizing a bivariate relationship involves generating an *estimate* of the direction and strength of the *X-Y* relationship. Second, in ANOVA and proportions models, it was important to find out, if possible, if the levels or groups of the predictor variable were *randomly arranged* or *selected*. In this section we follow up on those points because they are crucial to accurately interpreting and understanding a relationship in ways that go beyond bivariate statistics.

2.3.1   Generalizability

Data analysis is done in sample data, and any research question that can be addressed by the data can be answered *with certainty*—for the sample. For example, earlier in the chapter we noted that the mean difference between female and male NHANES survey participants' level of physical activity was 12 days per week, with females being slightly more active than males. We know that statistic is true for the sample of participants we analyzed, but the sample came from a much larger population of households. Often the sample is only a very small piece of the population. Furthermore, samples differ, and our sample (in this case, the 2011–12 NHANES survey sample) is just one of a very large number of possible samples that could have been drawn. Nevertheless, it is highly desirable to be able to generalize our sample findings to the population from which the sample came. **Generalizability** is the ability to apply sample findings, with at least some confidence, to the population from which the sample came (Fig. 2.1).

The diagram above shows the dynamic relationship between population and sample and the process of inferring population values, called **parameters**, from sample **estimates** of those values (i.e., statistics). Some inferential tasks, such as hypothesis testing and parameter estimation, are heavily *statistical* in nature. We will dive deeply into those statistical methods later in the book (Chap. 7 and beyond). For our present purposes, we are interested in *methodological* inference. In other words, what elements of the study design and method allow us to make better inferences about population parameters and characteristics? Let's focus on one element of study design—the sampling method.

Sampling is the process by which study participants (be they people, animals, or other observational units) are drawn into a study sample from the population in

Fig. 2.1 The population—
sample dynamic in
methodological inference

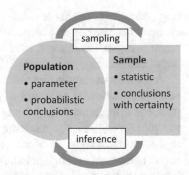

which they exist. There are two sampling methods that we need to be able to distinguish and, if possible, identify in the study documentation of the data we analyze. **Random sampling** uses some kind of random process (e.g., random numbers) to select participants from the population for the study sample. A random process ensures that every population member has an equal chance of being selected for the sample. **Convenience sampling** uses one of a variety of non-random methods to select participants for a study. Here are some examples of convenience sampling:

- Sending a sign-up sheet around in classes to get names and contact information of potential participants
- Sending a survey link to an organization's listserv
- Offering extra credit to students for participation in a study

The sampling method used in a study determines the generalizability of the study's findings. **Generalizability** refers to the extent to which we can apply an observed (i.e., sample-based) X-Y relationship to the population from which the sample came. Random, compared with convenience, samples are associated with increased generalizability. This is because random samples represent the population much better than convenience samples. In terms of population parameters (i.e., the true value of some population dimension) and characteristics (i.e., the true proportion of population subgroups), random samples provide a more accurate "picture" of the population and therefore we have more confidence that parameter estimates from random samples are more accurate than estimates from convenience samples.

We know that samples differ in their estimates; indeed, our study sample is only one of a great many possible samples we could have drawn from the population. With random sampling, that difference between sample estimates is assured to be random. That's a really important quality, and we'll learn more about it in Chap. 7. With convenience samples, on the other hand, differences in sample estimates are not random. Think about it: if you allow participants to "select" themselves into your study sample—because they want the extra credit, or because they're free at the time of the study, or for any other reason—your sample will be biased by the factors on which participants self-selected into the sample. The sample might over-represent a particular subpopulation (e.g., students) or a particular characteristic (e.g., need extra credit) or both. Those biases severely limit the generalizability of convenience sample estimates. Moreover, even if we were to assume that a convenience sample was representative of *some* population, we wouldn't know what that population is.

A note about sample size: although random samples always generate more accurate and trustworthy estimates of population parameters than do convenience samples, the sample needs to be large enough to ensure its representativeness. Small samples, even if they're randomly selected from the population, are likely to misrepresent characteristics of the population. What is a good sample size? There is broad agreement among statisticians that a minimally acceptable sample size for a random sample to approach representativeness is around 100. At the other end of the spectrum, most political polling and attitude research on large populations (e.g., American adults) is done with random samples of 1200–1500, and larger samples don't return much if any more accurate estimates.

2.3.2   Causation

Another study characteristic that has important implications for interpreting data analytic results is whether the study design allows us to conclude that an observed *X-Y* relationship reflects an underlying *causal* relationship or merely an *association* between the variables. We noted earlier in the chapter that for ANOVA and proportion model studies we should determine, if possible, if the predictor variable levels are randomly arranged or selected. How does that matter to inference? "Correlation does not imply causation" is an oft-repeated caveat about interpreting correlational research. To understand what the caveat means, let's consider the causal possibilities that exist in the relationship between two variables we know are substantially related: stress (*X*) and health (*Y*).

There are three causal possibilities that could produce a relationship, or a correlation, between two variables:

1. $X \longrightarrow Y$

2. $Y \longrightarrow X$

3. $X \longleftrightarrow Y$
       $\searrow \quad \nearrow$
          Z

In terms of our example, #1 shows the causal effect of stress on health, #2 shows the causal effect of health on stress—which is equally plausible and likely—and #3 represents all the "3rd variables" that could be associated with both greater stress and poorer health, such as personality or lifestyle variables. For example, poverty can be associated with both greater stress and poorer health.

So if we find a relationship between stress and health in our sample data, can we conclude that higher stress *caused* poorer health? As with generalizability, it's the design of the study that allows us to make causal inferences about an *X-Y* relationship or not. A study with high internal validity allows for more confidence causal inferences. **Internal validity** is the degree to which a study establishes that an observed change in *Y* is due to (i.e., caused by) *X* and *X* alone. What contributes to internal validity?—two things. First, a study that is designed so that the variables have *temporal order* (X occurs before Y) rules out causal possibility #2 above. If, for example, we measured stress in September and health in December, it is not possible for the later event to cause the earlier one. Second, a study that is designed to *control all alternative explanations* for the change in Y (in other words, all the Z variables), other than X, rules out causal possibility #3 above. The main design element for controlling alternative explanations is random assignment of participants to groups. **Random assignment** of participants to groups takes the study sample and uses a random process to allocate participants to one group or the other. Random assignment assures that the two groups of participants are equivalent on all (Z) variables except for X. The logic of experimentation states that if two groups of participants are the same at time 1 (assured by random assignment) and different at time 2 (the measurement of Y shows the groups differ), the change must have been caused by the predictor variable, which was the only thing allowed to vary between time 1 and 2. In terms of our stress and health example, imagine we randomly assigned half of our participants to a high-stress task condition (work on an unsolvable problem with a time limit) and the other half to a low-stress (work on an easy, solvable problem with no time limit) task condition. Then we measured participants' blood pressure (as a proxy measure for health) and find that the high-stress compared with the low-stress participants are substantially higher. That study would have high internal validity, because of the confidence we have that no other factor could have produced the change in blood pressure than the stressor. And now we see why this is relevant to ANOVA and proportions model studies—those are the models with a categorical X.

A study that randomly assigns sample participants to levels of the X variable is called an **experiment**. Any other non-random method of participant allocation to group, such as selecting participants who *already have* the levels of the predictor (e.g., people who work in low-stress and high-stress jobs), produces a **correlational study**. There are two types of correlational studies. Cross-sectional studies measure X and Y at the same time. Like typical survey studies, cross-sectional research cannot establish temporal ordering. Longitudinal studies measure X before Y but do not (or cannot) randomly assign participants to levels of the predictor. Longitudinal research establishes temporal ordering of X and Y but does not control alternative (Z) explanations. Finally, keep in mind that random assignment is independent of sampling. A study can have a convenience sample (low generalizability) but randomize the sample participants to levels of the predictor variable (high internal validity) *or* a random sample (high generalizability) that is used in a correlational study (low internal validity). These study design elements and their implications for interpreting data analytic results are summed up in Table 2.4.

Table 2.4 Study design elements and their relationship to generalizability and internal validity

Does the study…		
Randomly sample from the population of interest to form study sample?	Randomly assign sample participants to treatment/comparison groups?	Inferences allowed regarding generalizability and internal validity:
No	No	Findings don't generalize to the population Study cannot establish causal effect of X on Y
Yes	No	Findings can be generalized to the population Study cannot establish causal effect of X on Y
No	Yes	Findings don't generalize to the population Study can establish causal effect of X on Y
Yes	Yes	Findings can be generalized to the population Study can establish causal effect of X on Y

2.4 Summary

Because they are foundational to good data analysis, let's recap the big ideas from this chapter. Plus, we will need to return to these ideas often as we move through the rest of the book.

- **The statistics and plots we use to describe an *X-Y* relationship depend on the statistical model of the study**. Diagnosing the correct statistical model, in turn, depends on identifying the predictor and outcome variables and the measurement scale of each. This information must be retrieved from the study documentation. If that documentation is unavailable, the analyst must make assumptions and conduct the analysis using the statistical model made with those assumptions.
- **Interpreting data analytic results with regard to their generalizability requires knowledge of the study's sampling method**. Diagnosing the sampling method again involves digging it out of the study documentation. If the study used a random sample, the population from which it came (i.e., population of interest) is also helpful to incorporate into our interpretation of the results. If the study used a convenience sample, the particular method used to create the sample may provide clues both to a plausible population of interest and to the ways in sample bias may be related to the outcome.
- **Interpreting data analytic results with regard to their internal validity requires knowledge of the study's design**. If the groups comprising the predictor variable were randomly arranged (i.e., experimental design), the study is

capable of showing the causal effect of X on Y. If the predictor variable groups were selected (i.e., observational design), then the predictor variable is confounded with many other variables—any of which could account for the observed X-Y relationship—and the study cannot establish that X caused Y, even if they are related. Study design information must be retrieved from the study documentation. If that documentation is unavailable, the analyst must make assumptions and conduct the analysis under those assumptions.

Together these principles remind us that *how we analyze* data and *what we can conclude* from those data analytic findings are circumscribed by the particular research methods used in the original study. It is important that our data analysis pay attention to that context.

2.5 Exercises

1. Identify the measurement scale of the following variables:

    (a) Political affiliation (Dem, Rep, Indep)
    (b) Customer satisfaction rating (1–5 scale)
    (c) Temperature (degrees F)
    (d) Temperature (degrees Kelvin)
    (e) SAT score
    (f) GPA (4 pt. scale)
    (g) Restaurant review rating (5-star scale)
    (h) Height (cms)
    (i) Weight (kgs)
    (j) Assignment score in a class (1–10 scale)
    (k) HS class (fr, soph, jr, sr)
    (l) Education (years)
    (m) Handedness (left, right, ambidextrous)
    (n) Family type (1 parent/no sibs, 1 parent w/sibs, 2 parent/no sibs, 2 parent w/sibs)
    (o) Pulse (bpm)

2. Identify the predictor (X) and outcome (Y) variables and the statistical model (ANOVA, proportions, regression, or logistic) applying to each study description:

    (a) Do girls who are in first grade do better at fine motor skills test than boys who are in first grade?
    (b) Are students who take an SAT-prep course (vs those who don't) more likely to get accepted (or not accepted) to college?
    (c) Is number of extracurricular activities (sports + clubs + organizations) in high school related to college GPA?
    (d) Is number of extracurricular activities (sports + clubs + organizations) in the 1st year of college related to retention (returning or not for the 2nd year)?

(e) Are children from families with siblings (vs only children) more extraverted (measured by a test of extraversion)?

(f) Does father's education (in years) predict whether his children graduate from high school (or not)?

(g) Are left-hand dominant (vs right-handed) children more creative (measured by a creativity test)?

(h) Is handedness (left-hand vs right-hand dominant) in college students related to major (STEM vs non-STEM)?

(i) Are male vs female HS teachers more likely to teach science/math courses vs other subjects?

(j) Is body mass index (BMI) related to sleep quality (measure by sleep quality test)?

Chapter 3
Statistics and Data Analysis in an ANOVA Model

Covered in This Chapter

- Statistics and plots for ANOVA model data
- Methods for diagnosing influential values
- Interpreting group differences in an ANOVA model

3.1 Data Analysis in an ANOVA Model

In Chap. 1 we reviewed a set of fundamental statistical concepts and tools and used them to summarize the properties of a numeric variable. In Chap. 2 we learned that data analysis and interpretation are closely tied to design elements of the study that produced the data, including the statistical model inherent in the research question, how the study sample was created, and whether sample participants were randomly allocated to treatment or comparison groups in the study. Building on the foundation of Chaps. 1 and 2, starting in this chapter and continuing through Chap. 6, we cover statistics and data analytic methods for describing relationships between variables.

Research questions rarely involve just a single variable. Most research questions are interested in the effect of an independent or predictor (X) variable on a dependent or outcome (Y) variable, or in the relationship between X and Y variables. This chapter deals with ANOVA model data. Recall, an ANOVA model consists of categorical X and numeric (i.e., interval or ratio scale data) Y variables.

$$numeric\, Y = \beta_0 + \beta_1 categorical\, X + \varepsilon$$

In this chapter we will deal with ANOVA models in which X is a two-category variable. This is a very common statistical model, and some examples of ANOVA model

B. Blaine, *Introductory Applied Statistics*, https://doi.org/10.1007/978-3-031-27741-2_3

research questions are: "Do Kindergarten girls and boys differ on a test of math achievement?" or "Do smokers compared to non-smokers have higher blood pressure?"

Our primary goal for analyzing data in an ANOVA model is to describe the X-Y relationship with one or more statistics. We also want to learn what the data shows us about the X-Y relationship and to do that we extend and elaborate on the graphical tools learned in Chap. 1. Exploratory data analysis is essential for not getting trapped by our own (possibly biased) research questions as guides to relationships, so exploratory data analytic skills are as important to learn as methods for answering research questions.

3.2   Describing the X-Y Relationship in an ANOVA Model

As we laid out in Chap. 2, describing the X-Y relationship in an ANOVA model involves comparative statistics because we want to compare the two groups that make up the predictor variable on the outcome variable. Comparative statistics can take the form of a difference between statistics or a ratio of statistics; we look at examples of both below. Any statistic that captures an X-Y relationship is called an **effect size statistic**. An effect size statistic contains information about the direction and the size, or magnitude, of the effect[1].

Recall that there are two parameters in an ANOVA model: β_0, which is the average predicted Y score on the outcome when $X = 0$, and β_1, which is the estimate of the strength and direction of the X-Y relationship. Since our interest is almost always the X-Y relationship, our data analytic efforts are focused on β_1. Numerous statistics are available to capture the relationship in an ANOVA model. Below we cover location-based effect size statistics as well as some effect size statistics that are based on other concepts.

3.2.1   Location-Based Effect Size Statistics

The most common effect size statistics used to summarize the X-Y relationship in an ANOVA model are location-based statistics. These statistics describe how average or typical scores in group 1 differ from average or typical scores in group 2. Naturally, we use many of the same location statistics that we learned in Chap. 1 to form these comparative effect size statistics. The **mean difference** (MD, Formula 3.1) is simply the difference between the group means (for groups 1 and 2, respectively) on the outcome variable. The mean difference is in the units of the Y variable

---

[1] Note that we call statistics that describe an X-Y relationship effect size statistics, but that doesn't imply that there's a *causal* effect between the predictor and outcome. Remember, evidence for a causal relationship between x and y is determined by the study design (see Chap. 2).

measurement. Note that the order in which you subtract one group mean from the other determines the sign of the statistic and the sign indicates the direction of the *X-Y* effect or relationship. The ordering of the groups in the data, however, is often arbitrary, and so the analyst can arrange the mean difference (to be positive or nega-tive, as desired) to assist with interpretation and communication. Here's an exam-ple: Say we're comparing reading scores of 1st grade girls and boys, and our theory leads us to hypothesize that girls will have higher reading achievement than boys. If the girls' mean is indeed larger than the boys', we might want to calculate the mean difference by subtracting the boys' from the girls' mean, thus creating a positive mean difference statistic. This will tell us *how much* higher, on average, girls score than boys.

$$MD = mean(Y_1) - mean(Y_2) \tag{3.1}$$

Like the mean, the mean difference is a non-resistant effect size statistic. But since extreme scores can potentially exist in both groups' achievement data, they exert a more complex influence on the mean difference than we observed in Chap. 1 working with a single variable. For example, different patterns of outliers in the two groups' data can enhance or suppress the mean difference, or they can cancel their influences over the mean difference. In data with outliers or even in data that is skewed (i.e., asymmetrically distributed), it is good practice to generate a more resistant effect size statistic to describe the *X-Y* effect or relationship.

$$MD_{tr} = trimmed\ mean(Y_1) - trimmed\ mean(Y_2) \tag{3.2}$$

$$MedD = median(Y_1) - median(Y_2) \tag{3.3}$$

The **trimmed mean difference** (MD_{tr}, Formula 3.2) and the **median difference** (MedD, Formula 3.3) are robust alternatives to the mean difference. Their properties and interpretation reflect the respective statistics as covered in Chap. 1. A trimmed mean difference captures the mean difference after a set percentage of data is removed from the upper and lower ends of each group's distribution of values. More trimming yields a more resistant statistic, but also one that is less sufficient. The median difference captures the difference between the middle values of each group's distribution of values. The median difference, like the median, is unaffected by the presence of extreme scores but also has low sufficiency. Both statistics are inter-preted in the units of the Y variable measurement.

3.2.2 Other Effect Size Statistics

Although the mean difference is the most common statistic used to describe the *X-Y* relationship in an ANOVA model, we are not limited to location-based statistics. In fact, mindful of the opportunity to learn from data through exploratory analysis, we should look at the relationship through other statistical lenses. Toward that end, researchers are often interested in how groups differ on a measure of variability. That interest can simply be to show that groups' variances are *not* different. Some research questions, however, actually hypothesize effects of an intervention for the groups' variability. Here's an example: Say we are interested in the effect of online homework, compared to standard homework practice, for grade school math achievement because we believe instantaneous feedback to the student is better for learning. In a study students are randomly assigned to do their homework through a digital app which delivers immediate feedback, or through worksheets submitted to the teacher. We might expect that students in the experimental (digital) condition would be *more different from each other*, and thus have a greater variance, than students in the control (worksheet) condition. Why?—because it is not unusual for experimentally imposed treatments to be experienced differently by, and have different effects on, participants, resulting in different group variances. And that can help us discover potential explanations for those different experiences.

The **variance ratio** (VR, Formula 3.4) captures the variances of two groups of scores in an ANOVA model in ratio form. Again referring to the often arbitrary ordering of categories in a categorical X variable, creating the variance ratio so that it is greater than 1.0 (i.e., put the larger group variance over the smaller) makes the statistic a little easier to communicate and interpret. The variance ratio describes *how much greater* the larger group variance is than the smaller group variance. The units of the variance ratio are essentially percentage points, not units of the Y variable measurement. For example, VR = 1.5 means that the variance of the more variable group is 1.5 times, or 50%, larger than the variance of the less variable group. There are some rules of thumb that help analysts to decide when group variances are substantially different (e.g., VR > 3.0) and recommend further exploration as to why.

$$VR = \frac{var\left(Y_1\right)}{var\left(Y_2\right)}$$

$$(3.4)$$

Another statistical "lens" for describing an X-Y relationship in ANOVA model data is an effect size statistic that combines location and variability statistics from the groups. One of the drawbacks of the mean difference (MD) is that it doesn't take into account the variability of the two groups' data. To see the consequences of this, the figures below show two distributions of outcome variable data with the same mean difference but different group variability. The plot on the left shows two groups with a mean difference of 2 and each group having a standard deviation of 3. The plot on the right shows two groups with the same mean difference but much less variability within the groups. Which plot shows a clearer difference between groups? If you picked the plot on the right, you're correct! And that's because the mean difference captures only one property of the outcome variable. A treatment variable's effect can also be understood as its ability to *separate* participants in the two groups, or make their scores overlap less. So the group variability (in this case as measured by the standard deviations) turns out to be pretty important to evaluating the mean difference (Fig. 3.1).

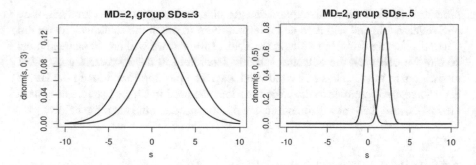

Fig. 3.1 Normal distributions with the same mean difference and different group standard deviation

Cohen's d (d, Formula 3.5) is a standardized mean difference statistic. The "standardization" is done by dividing the mean difference by the average of the groups' standard deviations, which produces a statistic that renders the mean difference in standard deviation units. Cohen's d allows us to distinguish between the two relationships in the plots above. On the left, where the two groups' data heavily overlap, Cohen's d = 0.67, which means that the difference between the means is 0.67 standard deviations. On the right, where there is little overlap in the groups' data distributions, Cohen's d = 4.0, indicating that the difference between the means is 4 standard deviations.

$$d = \frac{mean(Y_1) - mean(Y_2)}{\left(\dfrac{sd(Y_1) + sd(Y_2)}{2}\right)} \tag{3.5}$$

Cohen's *d* is a valuable effect size statistic for describing an *X-Y* relationship because it combines information about location (i.e., average effect of the treatment) with variability (i.e., the treatment's influence on group variability and overlap of scores). However, we cannot interpret the *X-Y* relationship in the *Y* variable measurement units, because Cohen's *d* converts them to standard deviation units. So whenever the *Y* measurement units are intuitive, such as pounds lost or blood pressure points lowered as a result of a diet or treatment, the mean (or trimmed mean) difference should be reported. Despite having nonintuitive units, Cohen's *d* allows analysts and researchers to compare treatment effects (or differences on *Y* associated with *X*) across subgroups, samples, and studies.

3.2.3   Plots

Graphical displays of ANOVA model data are important complementary tools to statistics for description and exploration, in part because they afford explicit visual comparison of data in the predictor variable groups. We will apply the same three

plots learned in Chap. 1 (histogram, density plot, and boxplot) to the analysis of an
X-Y relationship and will read and interpret them in much the same way as we did
when displaying data from a single variable. However, we will create paneled plots
so that we can view the outcome variable separately in the respective groups but
also compare the displays in a controlled way. The plotting functions in the lattice
R package are preferable to the basic functions we used in Chap. 1 for creating paneled plots, and these new R functions are demonstrated in the example below.

3.3   Data Analytic Example 1

To demonstrate the effect size statistics covered above, let's return to the 2011–2012
NHANES data used in the Chap. 1 example. In this example we are interested in the
relationship between gender (X) and systolic blood pressure (Y). All of the code in
the example below is in ASRRCh3.R.

The chunk of output below shows the following, in order:

- Read in the data file and assign it to an object name (dat).
- Generate means, 10% trimmed means, and medians for systolic blood pressure
  by gender (i.e., separately for female and male survey participants). This step
  allows you to see the group statistics before you compute the effect size statistics,
  in case you want to arrange the sign of the statistic for interpretation purposes as
  discussed above.
- Calculate the mean difference (MD) and the median difference (MedD).
- Calculate the variance ratio and Cohen's d. It is recommended that you calculate
  the VR so that it ends up being greater, rather than less, than 1.0. For that reason
  we first find the group variances and then compute the VR by dividing the larger
  variance by the smaller variance. Cohen's d can be calculated using Formula 3.5
  and the required descriptive statistics. The analysis below uses a function in the
  effsize package, so that package must be installed first. As we did with boxplot
  output in Chap. 1, we retrieve the statistic (d) from the function output with the
  element name. The sign of Cohen's d is determined by the R function computing
  d based on the ordering of gender categories in the X variable (i.e., first category
  mean minus second category mean). For reporting purposes it is common, however, to report Cohen's d as positive and convey the direction of the direction of
  the relationship with words.

```
library(NHANES)
dat<-NHANES[NHANES$SurveyYr=="2011_12",]

#find location statistic by group

tapply(dat$BPSys1,dat$Gender,mean,na.rm=TRUE)

##    female      male
## 117.7469 120.8369

tapply(dat$BPSys1,dat$Gender, function(x) mean(x,trim=0.1,na.rm=TRUE))

##    female      male
## 115.9683 119.6505

tapply(dat$BPSys1,dat$Gender,median,na.rm=TRUE)

## female    male
##    114     120

#use R to find group means and calculate MD and MedD
MD<-mean(dat$BPSys1[dat$Gender=="male"],na.rm=T)-
  mean(dat$BPSys1[dat$Gender=="female"],na.rm=T)
MD

## [1] 3.089946

#use R to find group medians and calculate MedD
MedD<-median(dat$BPSys1[dat$Gender=="male"],na.rm=T)-
  median(dat$BPSys1[dat$Gender=="female"],na.rm=T)
MedD

## [1] 6

#other statistics
#VR
tapply(dat$BPSys1,dat$Gender,var,na.rm=TRUE)

##    female      male
## 334.1236 279.1089

VR<-var(dat$BPSys1[dat$Gender=="female"],na.rm=T)/
  var(dat$BPSys1[dat$Gender=="male"],na.rm=T)
VR

## [1] 1.197108

#Cohen's d
library(effsize)
d<-cohen.d(dat$BPSys1,dat$Gender,na.rm=T)
d$estimate

## [1] -0.176488
```

What do these statistics tell us about the relationship between gender and systolic blood pressure? First, males compared to females have greater average, or typical, systolic blood pressure, with the effect size dependent upon the statistic (MD = 3.1 blood pressure points, MedD = 6 blood pressure points). As the more resistant statistic, the median difference is less influenced by any extreme values or outliers in the groups' data and might be the more accurate estimator of the "true" (i.e., in the population of participants) difference in systolic blood pressure. As a standardized mean difference statistic, Cohen's d (d = 0.18) tells us that systolic blood pressure is 0.18 standard deviations larger in male than in female study participants. Finally, the variance ratio (VR = 1.30) tells us that variability in systolic blood pressure is about 1.3 times greater in females than in males.

Let's now generate the graphical summaries of the relationship between gender and systolic blood pressure. The output chunks below show the following:

- Histograms of systolic blood pressure paneled by gender. As with the hist() function in Chap. 1, in the lattice histogram() function, the breaks argument controls the smoothing. In the code below, we specify an axis that ranges from 0 to 300 mpg, with bins of 10 points for optimal smoothing. You can compare these plots with more and less granular versions on your own.
- Density plots of systolic blood pressure paneled by gender, specifying the same level of smoothing as in the histograms.
- Boxplots of systolic blood pressure paneled by gender.

```
#plots for visualizing data distributions by groups
library(lattice)
#histogram
histogram(~BPSys1|Gender,data=dat,
          type="count",
          breaks=seq(0,300,by=10),
          layout=c(1,2))
```

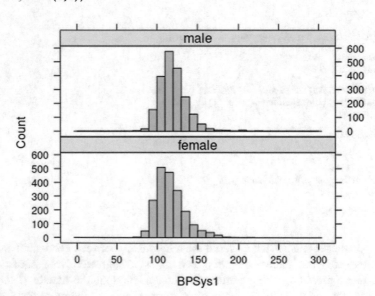

```
#density plot
densityplot(~BPSys1|Gender,data=dat,
            breaks=seq(0,300,by=10),
            plot.points="F",layout=c(1,2))
```

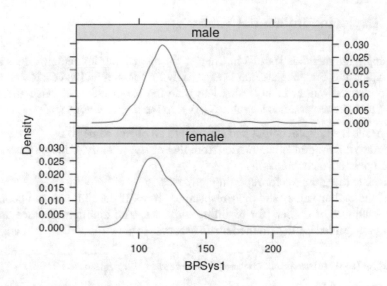

```
#boxplot
bwplot(Gender~BPSys1,data=dat)
```

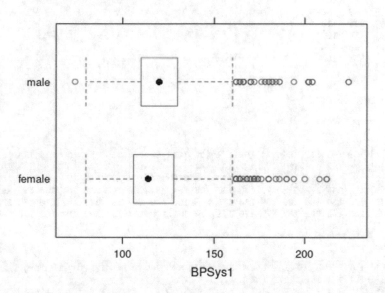

What do these plots tell us about the relationship between gender and systolic blood pressure? With regard to location, we notice in all three plots the general off-set of the distributions reflecting the higher typical or average blood pressure in male compared to female participants. With regard to variability, the boxplots reveal that blood pressure values are slightly less variable in males than in females (see IQRs). Finally, we notice that both groups have some extreme values and they are all in the positive end of their respective distributions (with the exception of a single very low value in the males' data). We explore these values in the next section.

3.4   Exploring Influential Values

Let's apply the methods learned in Chap. 1 for diagnosing influential values using
the boxplot rule, but generate output separately for females and males. Recall that a
boxplot-defined outlier is any value less than the lower adjacent value (LAV) or
greater than the upper adjacent value (UAV). In the output below, we:

- Generate the boxplot output and assign it to an object name (b).
- Retrieve the 5-number summaries from that object (mainly to identify the lower
  and upper adjacent values).
- Generate the group ns for reporting purposes.
- List the outlier values and their respective group. If desired, one can count the
  total number of outliers for reporting purposes. The highlighted cases are the
  boxplot-defined outliers in the female group.

```
b<-boxplot(Fuel.Information.City.mpg~transmission,plot=F,data=cars)

#5-number summaries and group ns
b$stats

##       [,1] [,2]
## [1,]   80   80
## [2,]  106  110
## [3,]  114  120
## [4,]  128  130
## [5,]  160  160

b$n

## [1] 2047 2060

#outlier values
b$out

##    [1] 168 168 162 164 174 168 172 162 162 164 164 164 164 164 174 174 162 184
##   [19] 208 162 168 164 172 172 172 164 172 168 174 212 212 164 190 190 162 194
##   [37] 164 164 180 180 180 162 170 170 170 164 162 172 164 174 200 176 162 162
##   [55] 162 186 172 172 172 172 166 168 164 164 162 162 180 164 162 166 166 164
##   [73] 170 164 180 180 204 204 178 166 166 178 162 186 186 182 170 170 224 224
##   [91] 202 202 202 202 202 170 172  74 166 170 194 184 182 182 162 176 164 164

#group assignments of outlier values
b$group

##    [1] 1 1 1 1 1 1 1 1 1 1 1 1 1 1 1 1 1 1 1 1 1 1 1 1 1 1 1 1 1 1 1 1 1 1 1 1 1
##   [38] 1 1 1 1 1 1 1 1 1 1 1 1 1 1 1 1 1 1 1 1 1 1 1 1 1 1 1 1 1 1 1 1 1 1 2 2 2 2 2 2 2 2
##   [75] 2 2 2 2 2 2 2 2 2 2 2 2 2 2 2 2 2 2 2 2 2 2 2 2 2 2 2 2 2 2 2 2 2 2

b$names

## [1] "female" "male"

length(b$out)

## [1] 108
```

What does this analysis tell us? First, as we observed from the boxplots earlier, with one exception in the males' data, all the outliers are in the upper blood pressure values. The UAV for both groups is 160, so any systolic blood pressure greater than 160 is defined as an outlier. Although there are a large number of outliers in each group, most of these outlier values can hardly be called extreme or unusual, and as such would not be expected to influence the mean much. This is reflected in small differences between the mean and 10% trimmed mean for each group. So overall we have no big concerns that outliers are exerting strong influence on non-robust summary statistics like the mean and SD.

3.5 Interpreting Group Differences from an ANOVA Model

In Chap. 2 we learned that what statistics can establish and say about an X-Y relationship depends on the sampling method used in the original study as well as on the experimental or correlational design of the study. Let's apply these principles (see Sect. 2.3 and Table 2.4) to the interpretation of group differences in ANOVA model data using the example from above in which we found a small relationship between gender and systolic blood pressure.

First let's deal with internal validity. As we learned in Chap. 2, internal validity is the degree to which a study establishes that an observed change in Y *is caused by X, and X alone*. Our data showed that the gender of the survey participant was indeed *related* to systolic blood pressure, but did being male versus female *cause* higher blood pressure, or can we only conclude that gender and blood pressure are related? To answer that, we need to look back (if possible) at the study design to determine if it was experimental or observational. If the predictor variable was experimentally arranged (by randomly assigning participants to one of two groups that vary on the X variable), then we have much more confidence that the observed difference in Y was caused by X. If however the X variable is based on groups of participants that already have the characteristic or not (as is the case with female and male survey participants), then the observed difference in the outcome variable could be explained by many other factors that are related to X. These "other factors" (often called confounding variables because they "confuse" the causal inference) are alternative explanations to X for the observed difference in Y. In our study there are many plausible alternative explanations or variables that relate to both gender and blood pressure (e.g., daily stress, physical activity, etc.). Since our study did not attempt to identify and control (eliminate or hold constant) those plausible confounding variables, we have little basis for concluding that gender caused the blood pressure difference. Evidence (or lack thereof) for causal inference should be reflected in the way we describe an observed group difference. In the example above, the data came from an observational study, because gender was selected rather than a randomly assigned status. Therefore, in summarizing the relationship, we use words like *association* or *relationship* to describe the X-Y relationship. In an experimental study, we may refer to the *effect* or *impact* of X on Y in our written summary.

Second, let's deal with generalizability. As we learned in Chap. 2, generalizability refers to the extent to which we can apply a study finding to the broader population from which the sample came. Was our observed relationship between gender and blood pressure limited to the particular sample of people in that study, or does the relationship generalize to the population from which the sample came? To answer that we need to look back at the study design to determine if random or convenience sampling was used to create the study sample. If the study documentation assures us that random sampling was used (which is the case in the NHANES study), then the lower (by 3 points) blood pressure in females than males observed in our study's sample can be generalized with some confidence to other random samples from the population, which is another way of saying "to the population." If some convenience sampling method was used, we have much less confidence that sample findings generalize to the population. Some reference to generalizability (or lack thereof) should appear in a written summary of the analysis of an X-Y relationship.

3.6   Data Analytic Example 2

In this example we analyze data from the gehan dataset in the MASS package. This dataset is from a clinical trial (which is another term for an experiment) in which a sample of leukemia patients was randomly assigned to a treatment or control condition (X); remission time in weeks was outcome (Y) variable. Clearly we have data from an ANOVA model (categorical X and numeric Y); our task is to summarize that X-Y relationship. Let's assume that the study used convenience sampling, meaning that study participants were patients who were available and willing to participate in the trial. They were not randomly selected from a population of leukemia patients. Many studies like this—experimental tests of clinical, educational, or other interventions—must do with small samples, whereas survey studies like our previous example often have the luxury of large samples. All of the code in the example below is in ASRRCh3.R.

The chunk of output below shows the following, in order:

- Library the MASS package to access the gehan dataset. The R function?, when followed by a dataset name, pulls up the documentation (i.e., sample and variable information, measurement units) that is available for all datasets that are included in R packages.
- Paneled histograms and boxplots of remission time by treatment group.
- Mean and median remission time by treatment group.
- Calculation of the mean difference, Cohen's d, and the variance ratio statistics.

```
library(MASS)

?gehan    #pulls up documentation for the dataset

#plots
histogram(~time|treat,data=gehan,
        type="count",
        breaks=seq(0,40,by=5),
        layout=c(1,2))
```

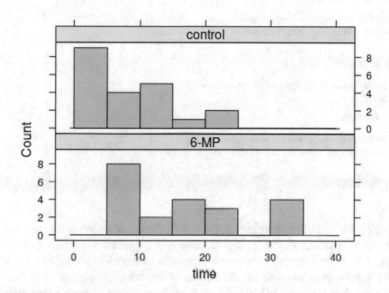

```
bwplot(treat~time,data=gehan)
```

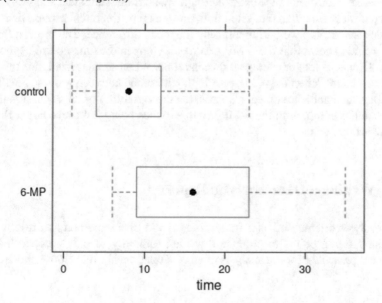

```
#statistics
tapply(gehan$time,gehan$treat,mean)

##      6-MP    control
## 17.095238  8.666667

tapply(gehan$time,gehan$treat,median)

##   6-MP control
##     16       8

MD<-mean(gehan$time[gehan$treat=="6-MP"],na.rm=T)-
  mean(gehan$time[gehan$treat=="control"],na.rm=T)
MD

## [1] 8.428571

d<-cohen.d(gehan$time,gehan$treat,na.rm=T)
d$estimate

## [1] 1.000909

VR<-var(gehan$time[gehan$treat=="6-MP"],na.rm=T)/
  var(gehan$time[gehan$treat=="control"],na.rm=T)
VR

## [1] 2.390211
```

Keeping in mind the sample size (N = 41), the plots show us a relationship in which patients in the 6-MP treatment have longer remission times than control patients. It's difficult to say how much longer from the plots alone, but in addition to a likely mean difference due to the treatment, the boxplots also show us two other things. One, there are no outliers in the data, and two, there is much more variability in the 6-MP compared to the control group. We start our statistical analysis by generating mean and median remission time (in weeks) by treatment group. Since these statistics are similar, we will report only the mean difference. The mean difference (MD = 8.43) shows that the 6-MP treatment, compared to the control, group had nearly 8.5 weeks longer average remission time. When standardized, this treatment effect results in Cohen's d = 1.0, or a 1 SD difference between groups. Finally, the data show us that the treatment also affected the variability of the remission time in the treatment group, with the 6-MP group variance being 2.4 times larger than the control group variance.

3.7 Writing a Data Analytic Report

Data analysis often culminates in a written report of the method and results. Here are four guiding points to produce a written summary of data analytic findings. These are presented for the analysis of an X-Y relationship in ANOVA model data.

1. **Research question**. State the research question, variables, and measurement details for the variables involved in the analysis. If it is known, state the sampling method (random or convenience) and design (experimental or observational) of the study from which the study came.
2. **Statistical summary**. It is conventional to report the mean and standard deviation on the outcome variable for each group, but robust location and variability statistics can be reported if they substantially differ from the mean and SD. If the sample size (N) and the group sizes (ns) weren't reported in #1, report those here. These six values can be organized into a table. Next, summarize the X-Y relationship with one or more effect size statistics and interpret the statistic's sign and magnitude with regard to the research question. By the way, there's no need to present every statistic you generated in your analysis, especially if they generally agree.
3. **Graphical summary**. Incorporate a graphical summary of the data (or two, if needed) to effectively communicate the X-Y relationship in the research question. These could be paneled histograms, density plots, boxplots, or some combination of those. Format your plots to have axis labels and other elements that help them communicate clearly (your instructor can help you learn how to do that in R). Create and report the plot or plots that communicate best.
4. **EDA**. Share any insights learned from exploratory data analysis. There may not be any insights with strong enough evidence to offer, but if there are, keep in mind that learning from data is an inductive exercise. The insights you offer are tentative and subject to later confirmation (or disconfirmation) so offer them with appropriate caution.

Transparency and reproducibility are essential characteristics of good data analysis. Even though the written summary is brief, it must be transparent. It should not mislead the reader as to the true nature of the findings or hide findings that are important for understanding the relationship. Reproducibility (i.e., the ability for your data analysis to be reproduced by someone else) is crucial for scientific credibility, so saving every step and procedure in your .R or .Rmd file makes the analysis recoverable and reproducible.

Following those guidelines, here is an example of a written summary of the analysis in Example 2:

Results

The effect of treatment on remission time (weeks) in a convenience sample (N = 42) of leukemia patients was investigated. Participants were randomly assigned to 6-MP or a control condition. Means and standard deviations for remission time are presented in Box Table 3.1.

Box Table 3.1 Remission time statistics by treatment group

Condition	n	Mean	SD
6-MP	21	17.1	10.0
Control	21	8.7	6.5

Patients in the 6-MP condition had longer remission time than control condition patients by an average of 8.4 weeks (Cohen's d = 1.0). The boxplots in Box Fig. 3.1 show that the median remission time for 6-MP (16 weeks) and control group (8 weeks) patients resulted in a similar relationship. The boxplots identified no outliers in the data. However, the IQRs in the plots suggest a possible treatment effect on the group variability. This was explored by generating a ratio of group variances (VR = 2.39). In other words, remission time variability was over twice as large in the treatment compared to the control group. This could indicate the presence of subgroups of leukemia patients who responded very differently to treatment.

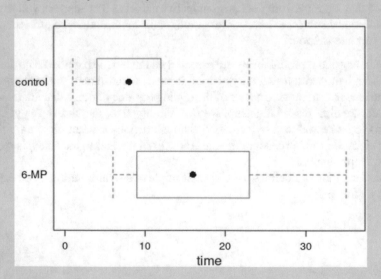

Box Fig. 3.1 Boxplots of remission time by treatment group

3.8   Problems

The problems below all use datasets in the MASS package. Use `library(MASS)`
to make the datasets available, and `?datasetname` *to see the documentation for*
each, which will provide variable names and measurement details.

1. In the `cats` dataset, summarize the relationship between `Sex` (x) and the `Bwt`
   (y) with the following statistics: mean difference, 10% trimmed mean difference,
   and median difference. Do the resistant and non-resistant statistics differ, and if
   so what does that suggest about the relationship. Why would we say "sex and
   bodyweight are related" rather than "sex has an effect on bodyweight"?
2. Using the `cats` dataset, find the range and IQR of bodyweight by levels of sex
   and calculate the variance ratio. Summarize what these say about the variability
   of bodyweight as it relates to the sex of the cat.
3. Using the `cats` dataset, generate histograms of bodyweight by sex of cat. What
   do the plots tell you about the distribution of bodyweight in male and female cats?
4. In the `Melanoma` dataset, compare the survival time (days) of males and females
   with the following statistics: mean difference, median difference, variance ratio,
   and the group's IQRs. Summarize the relationship between sex and survival
   time. Do you get a different picture of the relationship between sex and survival
   time from the mean difference compared with the median difference? Why?
5. Generate boxplots for the data in 3#4 and retrieve the 5-number summaries for
   each group. Are there outliers by the boxplot rule? In what group and what are
   the values?
6. Generate paneled density plots using the lattice `densityplot()` function to
   explore the relationship between sex and tumor thickness (mm). What do the
   density plots suggests about the distribution of tumor thickness for men and
   women with regard to differences in location, variability, modality, and
   extreme values?
7. Follow up the analysis in #6 and your hunches about the dimensions of tumor
   thickness distribution by generating statistics for each dimension. Interpret all
   the statistics you generate, noting if they are robust to the influence of extreme
   values or not.
8. Follow up the analysis in #6 by identifying outliers in the tumor thickness data,
   separately for men and women, using both the z-score and boxplot rules.
   Summarize your analysis.
9. In the `Melanoma` dataset do a full descriptive analysis of the relationship
   between ulcer condition (X) and melanoma survival time (Y). Generate appropriate statistics and plots, and write a summary reporting the key findings in your
   analysis.

Chapter 4
Statistics and Data Analysis in a Proportions Model

Covered in This Chapter
- Statistics and plots for proportions model data
- Contingency tables as data summary tools
- Interpreting group differences in a proportions model

4.1 Introduction

Recall from Chap. 2 (Table 2.3) that a proportions and ANOVA models have the same X side of the linear model equation—both models compare groups of participants on an outcome variable. However, a proportions model differs from an ANOVA model on the Y side of the equation in that we are now dealing with a categorical rather than a numeric outcome variable. Although categorical variables can have more than two categories, in this chapter we concern ourselves only with two-category outcome variables. Even with that constraint, many interesting and important outcomes take the form of two-category variables, such as a diagnostic test result (positive or negative) or college application decision (accepted or not accepted). Many questions about attitudes ("Do you approve of the President's performance?") or behavior ("Have you ever smoked marijuana?") are answered with a yes or no response, and these are also examples of categorical outcome variables. In this chapter we learn statistical and graphical tools and methods for analyzing proportions model data. The specific expression of the general linear model for a proportions model is below.

$$categorical\, Y = \beta_0 + \beta_1 categorical\, X + \varepsilon$$

Let's remind ourselves that in a proportions model, the data we want to summarize and analyze aren't "scores" but frequencies—each participant is counted, either in

B. Blaine, *Introductory Applied Statistics*, https://doi.org/10.1007/978-3-031-27741-2_4

one outcome category or the other. In the previous chapter, we learned that effect size statistics for ANOVA model analysis are built around a location statistic. For example, the mean was the basis for the both the mean difference and Cohen's d. With frequency data, the building block statistic is the **proportion**. A proportion is simply the number of cases in a category as a fraction of the total cases, in decimal form. Proportions and probabilities are very close concepts. In data from a study with categorical X and Y variables, we can think of a participant's "score" as a continuous measure of probability (of being in a particular category) that ranges from 0 to 1. Of course, participants are coded as *either* a 0 or a 1, but when we summarize a sample of participants, we do end up with a proportion—a number between 0 and 1 that describes the fraction of 1 s there are in a group of 0 s and 1 s. That proportion is also the probability that a given participant is in the "1" category.

Consider this example: imagine we get a random sample of ten individuals and ask them if they are right-handed or left-handed. Since we are interested in left-handedness, that will be our "presence" category (coded 1), and right-handedness becomes the "absence" category (coded 0). Our sample data are 0,0,1,0,0,0,1,0,0,0. If you find the mean of these "scores" (2/10 = 0.20), you have found the proportion of 1 s, or left-handers, in the sample. The 0.20 also is the probability that a randomly selected person will be left-handed. In studies with categorical X and Y variables, we are interested in comparing proportions of the target outcome (in our example, left-handedness) across levels of the X variable. For example, we might wonder if people who have a left-handed parent are more likely to be left-handed. Do you hear the phrase "more likely to be" in that question? That reflects the reality that with frequency data, we describe outcomes as more or less likely in one category of X than the other category. We will learn several new statistics that summarize a bivariate relationship between categorical X and Y in different ways, but they are all built on the basic ideas of the proportion and probability of a target outcome.

4.2 Contingency Tables

The **contingency** (or crosstab) **table** is a basic tool for describing proportions model data. Contingency tables allow us to see the distribution of frequencies across the four "cells" of the model, where each cell is a unique combination of predictor category and outcome category. To set up a contingency table for later analysis, we need to do two things:

1. We must designate the predictor and outcome variables. Applying the design concepts from Chap. 2: if these data came from an experimental study, the X variable would have been defined *by design*, meaning that the groups would have been randomly arranged, as when people are randomly assigned to drug or placebo conditions. When the predictor variable is not set by design (as is the case in observational studies), we must designate the X variable based on other considerations. Usually this is done by consulting the research question or by estab-

lishing via logic the most plausible causal model linking X and Y. This is an important step because we will format our contingency table with X in columns and Y in rows.

2. We need to give some thought to the "target" or "presence" outcome variable category. In the example above, we were interested in left-handedness, and therefore we coded the presence of left-handedness with 1 and right-handedness (as the absence of left-handedness) with 0. The category of interest is often communicated by the wording of the research question, but if that is not available, the analyst must consider other factors. As will become clear shortly, this decision affects the interpretation of effect size statistics we will compute to summarize the X-Y relationship.

The contingency table below shows a format in which the predictor is in columns, the outcome is in rows, with the target outcome category in the first row (Fig. 4.1). The letters a–d represent the frequencies for each cell of the table, as well as the marginal frequencies (i.e., column totals). Although a frequency table is the first step in summarizing proportions model data, raw frequencies prevent the comparison of groups on the outcome. Unless the marginal frequencies are the same (e.g., a + c = b + d), a more useful data analytic tool is a proportions table. A contingency table of proportions (Fig. 4.2) is generated from the formulas above and displays *conditional* proportions of the outcome category. Conditional proportions simply mean that the proportion of a group of participants in a particular outcome category is dependent (or conditioned) upon the group they are in.

$$p(presence \mid group1) = a / (a+c) \tag{4.1a}$$

$$p(absence \mid group1) = c / (a+c) \tag{4.1b}$$

$$p(presence \mid group2) = b / (b+d) \tag{4.1c}$$

$$p(absence \mid group2) = d / (b+d) \tag{4.1d}$$

Let's use these concepts in a simple example. Imagine we are interested in the relationship between gender (X) and handedness (Y) and we have data on these two variables from a sample of 25 people. The contingency tables below show the frequencies and conditional proportions of handedness associated with gender

Fig. 4.1 Contingency table

	X	
Y	Group 1	Group 2
Presence	a	b
Absence	c	d
	a+c	b+d

	X	
Y	**Group 1**	**Group 2**
Presence	*p*(presence|Group 1)	*p*(presence|Group 2)
Absence	*p*(absence|Group 1)	*p*(absence|Group 2)

Fig. 4.2 Contingency table of proportions

Fig. 4.3 Hypothetical handedness data

	gender	
handedness	female	male
left	3	3
right	7	12

	gender	
handedness	female	male
left	0.3	0.2
right	0.7	0.8

categories. Notice that if we tried to compare frequency data (i.e., comparing cells a and b in the frequency table in Fig. 4.3) we would come to the erroneous conclusion that left-handedness is equally rare in females and males. But since the marginal frequencies differ ($n_{female} = 10$, $n_{male} = 25$), the proportions of females and males, respectively, who are left-handed actually differ. The proportion of females who are left-handed is greater than the proportion of males who are left-handed. And because proportions and probabilities are equivalent, another way to say this is that left-handedness is more likely in females than males.

In our example above, we can plainly see that left-handedness is more likely in females than males, but we don't know how much more likely, nor do we have any statistical lenses for "seeing" that relationship. So now let's turn to the statistics that are used to estimate the X-Y relationship in a proportions model.

4.3 Statistics for Describing the *X*-*Y* Relationship in a Proportions Model

As was the case in an ANOVA model, effect size statistics for describing an X-Y relationship in a proportions model are also comparative statistics. Comparative statistics can take the form of a difference between statistics or a ratio of statistics, and we look at examples of both below.

Risk Difference You can think about the conditional proportion in a two-way contingency table as the "risk" of having that particular outcome status in that particular group. In our example, the risk of being left-handed was 0.30 for girls and 0.20 for boys. The difference between the two conditional risks is the **risk difference (RD)**. The risk difference is analogous to the mean difference statistic in Chap. 3. Inasmuch as risk values range from 0 to 1.0, the risk difference can range from 0 to 1.0 where RD = 0 would indicate no relationship between *X* and *Y*.

$$RD = p(presence \mid group1) - p(presence \mid group2) \qquad (4.2)$$

In the handedness example, the risk of being left-handed conditioned on being a girl is 0.30, whereas the risk of being left-handed conditioned on being a boy is 0.20. The RD (0.30–0.20 = 0.10) describes the magnitude of the relationship, and the sign shows the direction of the difference. How do we interpret the RD? We can say that the "risk" of being left-handed is 10 percentage points higher in girls than in boys (note that proportions can be converted to percentages by multiplying by 100). The RD statistic is a little more intuitive when the outcome category of interest is inherently "riskier" than the comparison category (e.g., positive cancer diagnosis), but the RD is a widely used statistic even with variables such as ours.

Number Needed to Treat Another statistic that helps us evaluate the magnitude of the *X-Y* relationship in a proportions model is the **Number Needed to Treat (NNT)**. The NNT is particularly useful in studies where the *X* variable is a treatment that is intended to lessen or prevent a risky outcome. The NNT expresses the risk difference as the number of participants that need to be treated to prevent the risky outcome in one participant. The NNT is the inverse of the risk difference (see Formula 4.3 and the contingency table in Fig. 4.1).

$$NNT = 1 / \left(\frac{a}{a+c}\right) - \left(\frac{b}{b+d}\right), \ or \ \frac{1}{RD} \qquad (4.3)$$

As an effect size statistic, lower NNTs indicate a more effective treatment. The perfect NNT is 1, which indicates that each person who is treated will produce one additional positive outcome. For a rule of thumb, NNTs less than 10 usually indicate effective treatments. However, interpreting an NNT statistic may depend on factors such as the cost of treatment and the side effects of treatment. Expensive treatments and/or treatments with side effects may need to deliver more benefit (i.e., lower NNT) to be regarded as effective.

Risk Ratio Once we understand the absolute difference between conditional risks (e.g., RD), it is straightforward to see that the **risk ratio (RR)** takes the same proportions and casts them as a ratio. The risk ratio quantifies how much greater or less is the risk of some outcome in group 1 than in group 2. An RR = 1.0 indicates no relationship between X and Y and is equivalent to RD = 0. RRs > 1.0 indicate greater risk, and RRs < 1.0 less risk, as you compare the categories of the X variable in order.

$$RR = p(presence \mid group1) / p(presence \mid group2) \tag{4.4}$$

In our example, the RR = 0.30/0.20 = 1.50. How do we interpret the risk ratio? We calculated our RR with ratio of girls' to boys' conditional risks of being left-handed, so a correct interpretation of this statistic would be that girls were 1.5 times as likely as boys to be left-handed, or that girls had a 50% greater risk than boys of being left-handed.

Odds Ratio The **odds ratio (OR)** is a similar statistic to the risk ratio but is based on conditional odds rather than conditional risks. But what are "odds"? The odds of an event are the ratio of the probability of occurrence to the probability of *nonoccurrence* of the event (rather than occurrence to *total* in the RR). For example, if it has rained on 5 out of the last 10 days in Rochester, the odds of it raining are p(rain) = 0.5/p(not rain) = 0.5, or 1.0. In other words, rain and no rain in Rochester are equally likely. An odds *ratio* is simply a ratio of the odds of the outcome in two categories of a predictor variable (see Formula 4.4). So, if it has rained 6 out of the last 10 days in Syracuse, the odds of it raining there are p(rain) = 0.6/p(not rain) = 0.4, or 1.5. We can then talk about the relationship between upstate NY city (*X*) and rain (*Y*) with a ratio of the odds, OR = Syracuse odds/Rochester odds = 1.5/1.0 = 1.5. So, the odds of it raining in Syracuse are 1.5 times higher than in Rochester. (Disclaimer: I made up the rain data. I have lived in both cities and they have equally pleasant weather.)

$$OR = \frac{p(presence \mid group1)}{p(absence \mid group1)} / \frac{p(presence \mid group2)}{p(absence \mid group2)} \tag{4.5}$$

Let's now calculate the OR in our handedness example, using the table of proportions in Fig. 4.3 or Formula 4.3.

$$OR = (p(LH|girl) / p(RH|girl)) / (p(LH|boy) / p(RH|boy))$$
$$= (.3 / .7) / (.2 / .8)$$
$$= .43 / .25$$
$$= 1.72$$

How do we interpret an odds ratio? We let our language reflect the fact that the OR is formed with the odds of the event in two different groups or conditions. So, for our example, we would say that odds of being left-handed (compared to being right-handed) are 1.72 times higher in girls than in boys.

4.4   Data Analytic Example 1

To demonstrate the effect size statistics covered above, let's use the ovarian dataset that is part of the survival package in R. Consulting the documentation for the dataset, we learn that the data is from a randomized trial, which is another way to say a study in which ovarian cancer patients are randomly assigned to one of two treatments (what we call below "standard" and "aggressive" treatment). The outcome variable of interest is whether residual disease is present (yes or no) at the end of the trial. Let's also assume that the study sample was not randomly selected from the population of people with ovarian cancer but was a convenience sample. In this example our research question is: What is the relationship between treatment type (X) and residual disease (Y)?

The chunk of output below shows the following data analytic operations, in order:

- Library the survival package to access the ovarian dataset.
- Create functions to calculate each of the statistics covered above from a table of proportions.
- Address two data wrangling tasks to make our contingency tables easier to read, and the statistics computed from them easier to interpret.

  1. Notice in the ovarian documentation (?ovarian) the treatment variable (rx) is coded 1, 2 but provides no information about what those codes mean. Consulting the original study (see documentation under References), we learn that 1 is the standard or typical treatment condition and 2 is the condition where standard treatment is enhanced with other drugs, or what we're calling "aggressive" treatment. The first line of code under #data wrangling below specifies new labels for those categories and assigns those to a new variable name (treatmt).
  2. We always want to arrange a contingency table with the "risky" outcome in first row. This is done so the statistics estimating the X-Y relationship speak to the relationship between the predictor and the "presence" (not the absence) of the outcome. Again consulting the documentation, we see that the presence of residual illness is the "risky" category in this case. The second line of code under #data wrangling below reorders the levels of the original variable (resid. ds) so that the *presence of illness* will end up in row 1 of the contingency table. We also specify new labels for those categories (yes and no).

- Generate a frequency table, for general descriptive purposes, followed by a table of conditional proportions. Set up these tables with X in columns and Y in rows, as shown. Notice three things about the line of code under #proportions table. First, the prop.table() function requires a table object, which is why the table() function is passed to that function. Second, the argument margins = 2 creates a table in which the conditional proportions are in columns, not rows. Third, the prop.table() function output is saved to an object (t) which makes the proportions in the table easier to pass to the statistical functions.
- With the table of conditional proportions created and saved, we pass that table object to the statistical functions created earlier.

```
library(survival)


#create functions for later use
#each takes a prop.table object
RD<-function(tab) {
  rd=tab[1,1]-tab[1,2]
  print(rd)
}

NNT<-function(tab) {
  nnt=1/((tab[1,1]/(tab[1,1]+tab[2,1]))-(tab[1,2]/(tab[1,2]+tab[2,2])))
  print(nnt)
}

RR<-function(tab){
  rr=tab[1,1]/tab[1,2]
  print(rr)
}

OR<-function(tab){
  or=(tab[1,1]/tab[2,1])/(tab[1,2]/tab[2,2])
  print(or)
}

#data wrangling
ovarian$treatmt=factor(ovarian$rx,level=c(1,2),labels=c("standard","aggressive"))
ovarian$residual.disease=factor(ovarian$resid.ds,level=c(2,1),labels=c("yes","no"))


#create contingency tables
#frequency table
table(ovarian$residual.disease,ovarian$treatmt)

##
##        standard aggressive
##   yes         8          7
##   no          5          6

#proportions table
t<-prop.table(table(ovarian$residual.disease,ovarian$treatmt),margin = 2)
t
##
##        standard aggressive
##   yes 0.6153846  0.5384615
##   no  0.3846154  0.4615385

#statistics
RD(t)

## [1] 0.07692308

NNT(t)

## [1] 13

RR(t)

## [1] 1.142857

OR(t)

## [1] 1.371429
```

Let's interpret these statistics. The RD tells us that the risk of having residual illness is 7.6 percentage points higher after standard, compared to the aggressive, treatment. This difference might be seen as small, given that *overall* risk of residual illness following treatment seems high (61.5% and 53.8% of patients in standard and aggressive treatment, respectively). The NNT is 13, meaning that 13 people need to receive treatment before recurrence of illness is prevented in one patient. Both of these statistics point to a not very effective treatment. This treatment effect corresponds to a risk ratio of 1.14, which means that the presence of residual illness is 1.14 times, or 14%, more likely after standard, compared to aggressive, treatment. The odds of having residual illness after standard treatment are 1.37 times higher than the odds of having residual illness after aggressive treatment.

Finally, what can we conclude about causation and generalizability from these findings? Notice that I used the word "effect" just above when referring to the *X-Y* relationship. Because our data come from a randomized trial—where patients were randomly assigned to treatment condition—we have a strong basis for concluding that the observed relationship is causal. Because of high internal validity, then, we can conclude that aggressive compared to standard treatment *caused* the reduction in residual illness. However, if we assume that the data are from convenience sample of patients with ovarian cancer, these findings might not generalize to the population, and thus we have low generalizability.

4.5   Plots for Describing the *X-Y* Relationship in a Proportions Model

The **barplot** and the **pie chart** are among the most commonly used plots for frequency or proportion data. Below are examples of each displaying proportions for a four-category variable (see Fig. 4.4). One of the serious limitations of the pie chart for displaying proportions is that the areas of pie wedges (which is the graphical representation of the category proportion) are imprecise and hard to compare. In a barplot the height of a bar represents the category proportion. Looking at the pie

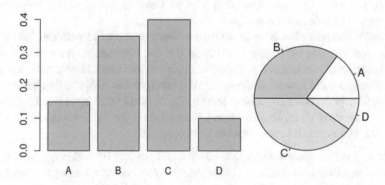

Fig. 4.4 Barplot and pie chart of hypothetical data

chart below, it is difficult to determine whether category B or C had the higher proportion. The barplot affords much more precision in determining the proportion in any particular category as well as comparing proportions across categories. It is relatively easy to "read" the barplot with regard to the proportions in categories B and C by finding where the top of the bar is on the X axis scale. Finally, the limitations of the pie chart are magnified when we want to display contingency table data. For displaying proportions model data, the **grouped barplot** is an ideal tool. It provides a readable graphical summary of conditional proportions and partners well with the summary statistics covered in the previous section. We will generate a barplot in the next analytic example.

4.6   Data Analytic Example 2

For this example we use the gbsg dataset that is part of the survival package in R. Consulting the documentation for the dataset, we learn that the data is from a 5-year study of breast cancer patients. Let's assume that the study sample is a convenience sample. In this example our research question is: What is the relationship between menopausal status (X) and disease remission (Y)?

Making sure that as in Example 1 the survival package is been libraried and we have created the statistical functions we want to use, the chunk of output below shows the following data analytic operations, in order:

- Address two data wrangling tasks to make our contingency tables easier to read, and the statistics computed from them easier to interpret.

  1. The gbsg dataset documentation identifies what the 0, 1 codes for the menopause status variable mean. Nevertheless, we need to attach labels to the 0 and 1 codes so that our contingency table (and, in turn, barplot) is appropriately labeled and readable. (I also reordered the levels of the variable so postmenopausal status is in column 1 of the contingency table. The ordering of the X variable categories in arranging the contingency table is the analyst's decision and your instructor can show you what would happen if we left the variable in 0, 1 order.) This reordered and labeled variable is assigned those to a new variable name (menopause.status).
  2. As in Example 1, we try to arrange a contingency table with the "risky" outcome in first row. Again consulting the documentation, we see that cancer recurrence or death is the "risky" category in this case. The second data wrangling operation reorders the levels of the original variable so that the *presence of illness* (in this case cancer recurrence or death) will end up in row 1 of the contingency table. We also specify new labels for those categories and pass those changes to a new variable (illness.status).

- Generate a frequency table followed by a table of conditional proportions.
- Calculate effect size statistics for the relationship between menopause status and illness status.
- Generate a grouped boxplot of the conditional proportions.

```
library(survival)

gbsg$menopause.status=factor(gbsg$meno,level=c(1,0),labels=c("post","pre"))

gbsg$illness.status=factor(gbsg$status,level=c(1,0),labels=c("recurrence/death","no
recurrence"))


#create contingency tables
#frequency table
tab<-table(gbsg$illness.status,gbsg$menopause.status)

#proportions table
t<-prop.table(tab,margin = 2)
t

##
##                          post          pre
##    recurrence/death 0.4545455 0.4103448
##    no recurrence    0.5454545 0.5896552

#statistics
RD(t)

## [1] 0.04420063

RR(t)

## [1] 1.107716

OR(t)

## [1] 1.197479

barplot(t,
        ylim=c(0,1),
        beside=T,
        legend=rownames(tab),args.legend=list(x="topleft"),
        main = "Menopause status and 5-year survival
        of breast cancer patients")
```

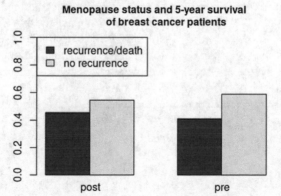

Based on this analysis, how do we describe the relationship between menopausal status and recurrence of illness? There is a higher probability of illness recurrence or death in the postmenopausal compared to the premenopausal patients, but the relationship is small. The risk of recurrence is only 1.1 times (or 10%) more likely (and the odds of cancer recurrence or death are only about 1.2 times higher) in the postmenopausal than in the premenopausal group. This small relationship is reflected in the barplot, which shows conditional probabilities of illness recurrence or death separately for the predictor variable groups. We didn't compute the NNT in this example because of the observational study design (i.e., menopausal status was selected, not randomly assigned). As discussed in Chap. 2, observational designs have low internal validity, which means that we have little basis for concluding that menopausal status caused the slightly greater risk of illness recurrence. Finally, if the sample of patients we analyzed was not randomly selected from the population of breast cancer patients, then our findings have low generalizability to that (or some other) population.

4.7   Writing Up a Descriptive Analysis

The written summary below used the data and analysis from the first data analytic example above. Our written summary goals are largely the same as we have applied them in previous Chap. 3, although a categorical outcome variable doesn't have "properties" to summarize. In addition, remember that cell frequencies and proportions are descriptive statistics, so it is appropriate to include either a table or a plot in your summary. Two additional guidelines apply if you include a table or a plot in a written summary. One, give the table or figure a title and number (tables are called "table," whereas plots are called "figures") and, two, refer to the table or plot in your write-up.

Results

A convenience sample (N = 26) of ovarian cancer patients was randomly assigned to standard or aggressive treatment. The outcome variable was the presence of disease at the end of the trial. The proportions of patients by condition with residual illness at the end of the trial are displayed in Box Table 4.1. The risk ratio shows that the risk of having residual illness was 1.14 times higher in the standard, compared to the aggressive, treatment condition. The NNT was 13, indicating that 13 people would need to receive treatment before recurrence of illness is prevented in one patient. Both of these statistics suggest that aggressive treatment is not much more effective than standard treatment for eliminating disease. Due to the experimental nature of the study design, however, the small reduction in risk of residual illness can be assumed to be caused by the aggressive treatment.

Box Table 4.1 Proportion of patients in standard vs aggressive treatment conditions with residual disease at the end of the trail

	Treatment condition	
Residual disease	Standard	Aggressive
Yes	0.62	0.54
No	0.38	0.46

4.8 Problems

The problems below all use datasets in the MASS package. Use `library(MASS)` *to make the datasets available, and*? `datasetname` *to see the documentation for each, which will provide variable names and measurement details.*

1. In the `survey` dataset, create a frequency table to explore the relationship between `Sex` (X) and the `W.Hnd` (Y). Generate a table of proportions and interpret the conditional proportions of being left-handed for females and males, respectively. Calculate and interpret the risk difference, risk ratio, and odds ratio statistics.

2. Generate a frequency table of the `Fold` variable. Recode that three-category variable into a new variable that drops the "neither" category using the line of code below (note that the recode function is in the car package).

```
survey$fold=recode(survey$Fold,"'Neither'=NA;'else'")
```

   Create a frequency table with `Sex` as the column variable and the new variable above as the row variable. Generate conditional proportions of the occurrence (left on right arm-fold) and interpret. Summarize the relationship between gender and arm-folding pattern with RD, RR, and OR statistics. What do they tell you about the relationship?

3. In the `Melanoma` dataset, generate a frequency table to explore the relationship between `sex` (x) and `ulcer` (y). It will be easier to interpret and use the table if the variable categories are renamed from 0 and 1 to the labels used in the code book (see example below). Summarize the relationship with RR and OR statistics.

```
Melanoma$ulcer_f=factor(Melanoma$ulcer,level=c(0,1),labels=c("presence
","absence"))
```

4. Explore the relationship between sex and survival status. First recode `status` to a two-category (dead/alive) variable and label its categories appropriately. The code below will help you; make sure you can explain what the recode statement is doing to the original variable.

```
Melanoma$status=recode(Melanoma$status,"3=1;'else'")
Melanoma$status=factor(Melanoma$status,level=c(1,2),labels=c("dead","
alive"))
```

   Generate a table of conditional proportions of dying (the occurrence of interest) for female and male patients, respectively. Summarize the relationship with RD, RR, and OR statistics. What do they tell you about the relationship?

5. Analyze the same outcome as in #4 but use ulcer as a predictor.

6. For either #4 or #5, write a summary of your analysis and include a barplot of the conditional proportions.

Chapter 5
Statistics and Data Analysis in a Regression Model

Covered in This Chapter
- Statistics and plots for regression model data
- Scatterplots as data summary tools
- Residuals analysis
- Interpreting correlations

5.1 Data Analysis in a Regression Model

We make an important transition in this chapter from the focus of the previous two chapters. In Chaps. 3 and 4, we analyzed bivariate relationships of categorical predictor variables with outcome variables. Many research questions however have predictor variables that are inherently numeric, such as:

- What is the relationship between years of education (X) and annual income (Y)?
- How is SAT score (X) related to college acceptance (Y, accepted/not accepted)?

Obviously, these questions don't involve a comparison. Instead of people having one of just two possible values on a categorical X variable, we now have data in which each participant can have any score on a numeric X variable. A regression model consists of numeric X and numeric Y variables, and in this chapter, we learn the statistics and plots that are useful for analyzing regression model data.

$$\text{numeric } Y = \beta_0 + \beta_1 \text{numeric } X + \varepsilon$$

In Chaps. 3 (ANOVA model) and 4 (Proportions model), our main statistical tools for describing bivariate relationships were *difference* statistics. Categorical predictor variables created groups of participants, and comparing those groups on the outcome variable with statistics like the mean difference and the risk difference

© The Author(s), under exclusive license to Springer Nature Switzerland AG 2023
B. Blaine, *Introductory Applied Statistics*,
https://doi.org/10.1007/978-3-031-27741-2_5

was an intuitive analytic approach. However, difference statistics won't work to describe an *X-Y* relationship in a regression model. Correlational statistics are a family of statistics that describe how two numeric variables *co-relate*, or in other words, how much *Y* changes in relation to a unit change in *X*. It turns out that many correlational statistics have their roots in the mean and variance, so in concept they are similar to statistics we covered in Chap. 3.

In this chapter we cover two statistical methods for analyzing data from studies with numeric *X* and *Y* variables: regression and correlation. Regression analysis is a powerful framework for analyzing all kinds of data and is far more informative than correlational analysis, so we put that front and center. We also take up another form of regression in Chap. 6, so the principles are important to establish here. Later in the chapter, we cover the family of correlation coefficients that are inherent to correlational analysis.

5.2   Logic of Linear Regression

In Chap. 3, we described how scores on a numeric *Y* variable related to a categorical change in *X*. Let's look at ANOVA model data through the lens of a regression model using data from the Early Childhood Longitudinal Study (ECLS, www.nces.ed.gov/ecls). The ECLS is a survey of children's knowledge, skills, and development from birth through elementary school. Figure 5.1 shows data from a small random sample of kindergarten students in which students' math test scores are plotted at each level of school type (public and private) in a plot called a scatterplot. Recall that the mean difference was one of the statistics we used to describe the *X-Y*

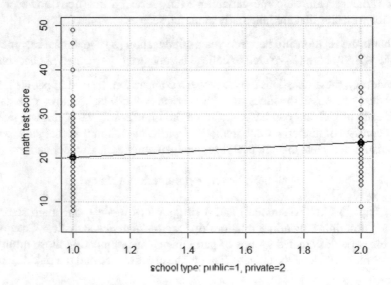

Fig. 5.1 Relationship between school type and kindergarten math achievement

relationship in an ANOVA model. The group means (public = 20.2, private = 23.6) are plotted with the solid black dots, and the mean difference is 3.4 points.

In Chap. 1 we learned that the mean is the balance point of the sample scores, in that the "weight" of the scores above and below the mean balance. Let's now move that concept to a bivariate scenario. If we fit a line to the data above such that the line is required to achieve a balance of the "weight" of the points above and below it in both groups, that line will go through the means of the two groups of scores (as shown in Fig. 5.1). That line is a regression line. Regression is a method for describing a relationship between a numeric X and a numeric Y variable using a linear function. In this example, we're pretending that public vs private school is a continuous numeric variable with, as it happens, only two points on its "continuous" scale.

Now, consider the slope of this regression line. The slope of a line is the average change in Y associated with a 1-unit increase in X. In our simple example, a 1-unit increase in X is a "movement" from public (X = 1) to private (X = 2) school. What happens to average math scores as you move from public to private school?—they increase by the mean difference, or 3.4 points. Although school type is not a numeric predictor, this example shows the essence of regression analysis. Regression analysis involves using a line to summarize a bivariate relationship (between numeric X and Y variables) by requiring that the line find the optimum balance between the weight of the points above it and below it. The line that achieves that optimum balance will have an equation, with particular slope and intercept values.

With those foundational ideas in place, let's learn how least squares regression analysis works and what statistics it produces to analyze relationships between numeric X and Y variables.

5.3 Least Squares Regression

In the above example, we saw that the regression line that achieved an optimum balance of the weight of points above and below it ran through the means of the two groups. However, when X is a numeric variable, we don't have discrete groups of participants; participants are merely defined by their scores on X. Even if we thought of each X score as a "group" of participants, it is highly unlikely that a straight line would run through all the conditional means of Y at each X value. To see that, the scatterplot below shows the relationship between kindergarten reading score (X) and first grade reading score (Y). The bold points show the conditional mean first grade reading score for five "groups" of students based on their X score. These are conditional means because they are the mean Y score *conditioned upon* a student having a particular value of X. Notice that the regression line (red dotted line) does not go through each conditional Y mean, but it is informed by them in the sense that those five "group" means are balance points for Y data at that particular X (Fig. 5.2).

Least squares regression is a method for fitting a line to a scatterplot that results in the best "fit," meaning that the *overall* balancing of points' weight above and below

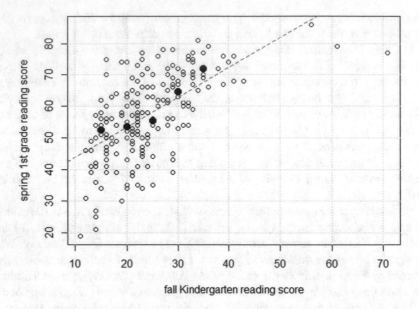

Fig. 5.2 Conditional means of Y for $X = 15, 20, 25, 30$, and 35

the line is achieved. Let's learn how that is done in statistical terms by creating a very simple dataset of five cases, which are plotted in Fig. 5.3 Plot 1 below. Now, for a moment, pretend that you don't have the X scores and I ask you to predict someone's Y score. Without any predictor (X) variable, your best prediction will be the mean of Y (that's reasonable: if Y scores are normally distributed, the mean would be the center or median value and therefore your best guess as to anyone's value).

Plot 2 in Fig. 5.3 shows for a particular case ($X = 8$) the mean-based predicted Y score (5.8) and the actual Y score (9). The difference between what you *predict* for someone's Y score and their *actual Y* score (symbolized $\hat{y} - y$) is an important quantity in regression because it captures—when those scores are squared and summed over all the cases—the total Y variance that is available to explain with a predictor variable. You can also think of that quantity ($\hat{y} - y = 5.8 - 9 = -3.2$) as an "error of prediction" because our mean-based prediction misses the actual value by that much.

Now let's use a predictor variable to predict that person's score on the outcome. Two principles apply in regression analysis: One, if there is an X-Y relationship, X will generate a better prediction of Y than the mere Y mean would. And two, the stronger the X-Y relationship, the better the prediction. Plot 3 in Fig. 5.3 shows these principles for our target ($X = 8$) observation. Because X is correlated with Y (higher X scores are associated with higher Y scores), X delivers a predicted Y value that is closer to the actual Y value than we achieved by using the Y mean.

Regression analysis partitions the total deviation (symbolized $y - \bar{y}$) into two pieces. The *regression score* ($\hat{y} - \bar{y}$, line b in Plot 3) is the difference between the predicted y score and the mean of y. You can also think of this score as the improvement in our prediction using X compared to using $\bar{y}$. The *residual score* ($y - \hat{y}$,

Fig. 5.3 Components of
least squares regression

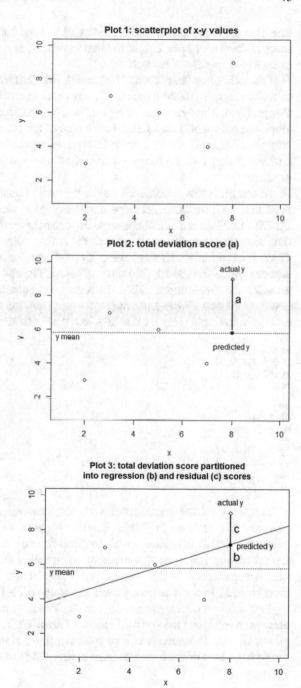

line c) is the difference between the actual y score and our predicted y. A residual score is the error of prediction; in other words, what's "left over" in the total deviation after we predict Y using X.

The optimal or "best fitting" regression line will minimize errors of prediction, or residual scores. Residual scores always sum to zero, but squaring them eliminates that problem. The *least squares regression line*, therefore, is the line (with particular slope and intercept) that minimizes the sum of the squared residual scores. Another way to understand it is that the least squares regression line finds the relationship between X and Y that results in the smallest *average* (over the range of X) error of prediction.

To complete this lesson, let's do the regression analysis for our $N = 5$ dataset in R so that we have the exact slope and intercept of the regression line (in Fig. 5.3, Plot 3) and so we can talk about some important regression statistics. We will do full data analytic examples, explaining all the R functions involved, in the next section. For this example we simply create X and Y data vectors and use them in a least squares regression model; the output is below. The slope and intercept form a linear equation for predicting y, called the regression equation, which in this example is $\hat{y} = 3.49 + 0.46x$. We can use the regression equation to predict Y from some value of X by simply plugging a value of x into the 0.46x term and finding $\hat{y}$.

```
x=c(2,3,5,8,7)
y=c(3,7,6,9,4)
reg=lm(y~x)
reg

##
## Call:
## lm(formula = y ~ x)
##
## Coefficients:
## (Intercept)            x
##      3.4923       0.4615
```

The slope of the regression line is the *unstandardized regression coefficient (beta, or β_1)*. It is an important statistic because it summarizes the relationship between X and Y. It's called unstandardized because the coefficient is understood in terms of X and Y units of measurement. Interpreting the coefficient follows the logic of a linear slope: it is the average predicted change in Y *(in Y units)* for a 1-unit increase in X. In our example, $\hat{y}$ increases 0.46 points for each 1 point increase in X.

Two more statistics are useful for describing a bivariate relationship through regression analysis: the *residual standard error (RSE)* and the *coefficient of determination (r^2)*. To understand how intuitive these statistics are, let's look at them through the lens of the formulas used to calculate them from data.

$$SS_{residual} = \sum (y_i - \hat{y}_i)^2 \tag{5.1a}$$

$$SS_{total} = \Sigma \left(y_i - \bar{y}_i \right)^2 \tag{5.1b}$$

$$RSE = \sqrt{\frac{SS_{residual}}{N-2}} \tag{5.2}$$

$$R^2 = \frac{SS_{total} - SS_{residual}}{SS_{total}} \tag{5.3}$$

Formulas 5.1a and 5.1b show the component quantities of RSE and r^2. First, residual scores (line C in Plot 3 above) are squared and summed up across all N participants to produce a sum of squared residuals. Similarly, deviation scores (line A in Plot 2 above) are squared and summed over N participants to produce sum of squared deviation scores, the total variance in the Y data. The RSE formula simply averages the $SS_{residual}$ scores (dividing by N-2) and then returns the result to the original measurement metric with the square root operation. Why do we average with N-2 instead of N? Because we're averaging residual scores, and to have those, you need to have a line, and to have a line, you need to "give up" two observations (because 2 points determine a line). The r^2 formula is simply regression variance (total − residual = regression; see Plot 3) divided by total variance.

The RSE is the average error of prediction in predicting Y from X. Remember that the least squares regression line is fit to the data such that the average residual score (or error of prediction) is as small as possible, and the RSE tells us what that error is. For that reason we refer to the RSE as a model fit statistic. The coefficient of determination (or r^2) is another model fit statistic and is the proportion (or multiplied by 100, percentage) of total Y variance that is explained by (or proportion of total error that is reduced by) the predictor variable.

To sum up what we have learned thus far:

- Regression analysis uses a smooth (linear) function to describe a relationship between a numeric X and a numeric Y.
- The regression line's slope and intercept are determined by the least squares method; the slope and intercept that minimize the sum of squared residual scores is the least squares or "best fit" line.
- The regression coefficient (β_1) describes the X-Y relationship in terms of average change in Y (in Y units) for a 1-unit increase in X.
- RSE is the average error of prediction when using X to predict Y.
- r^2 is the proportion of total variance in Y that is due to or "explained by" x.

5.4   Data Analytic Example 1

Let's do an example of least squares regression analysis using the 2011–2012 NHANES data in R and focus on the subpopulation of female smokers. In this example we analyze the relationship between physical health (X) and mental health

(*Y*) in the female smoker population. Checking the NHANES documentation, we see that the variables are measured as the participant's report of the number of days per month that they experience poor physical and mental health, respectively. As we have learned, it is important that you (a) designate *X* and *Y* variables for regression analysis and (b) do so with some awareness of their plausible causal connections. When variables are temporally ordered (i.e., *X* precedes *Y* naturally), the *X-Y* designation is straightforward. In this example, however, there is no temporal ordering, so we arbitrarily designate physical health as *X*, mindful that mental and physical health are probably dynamically related with regard to causation (i.e., either could plausibly cause the other).

We start by setting up a data frame of the 2011–2012 data subsetted to include only data of females who smoke. The analysis begins with a scatterplot with the least square regression line plotted on it, then we fit the regression model and retrieve from it the descriptive statistics presented above, and finally we use the regression equation to make some predictions. Let's do the analysis all at once and then work through the interpretations and implications of the results.

```
library(NHANES)
dat<-NHANES[NHANES$SurveyYr=="2011_12",]
#subset rows by levels of 2 factors
dat<-dat[which(dat$Gender=="female" & dat$SmokeNow=="Yes"),]

#scatterplot w/regression line
library(lattice)
xyplot(DaysMentHlthBad ~ DaysPhysHlthBad, data=dat,
       grid=TRUE,
       type=c("p", "r"),       #p=scatterplot, r=regression line
       col.line="red")
```

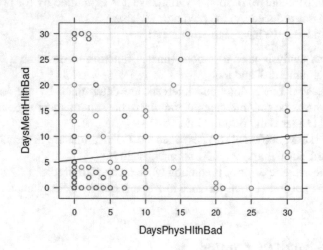

```
#do least squares regression and extract statistics from lm object
#lm() takes an y~x formula statement
reg=lm(DaysMentHlthBad ~ DaysPhysHlthBad, data=dat)

reg$coefficients     #intercept and slope

##      (Intercept) DaysPhysHlthBad
##         5.446950        0.155741

summary(reg)$sigma     #RSE

## [1] 9.04076

summary(reg)$r.squared   #R squared

## [1] 0.02331742

#predict from model
x=data.frame(DaysPhysHlthBad=c(0,10,20))
predict.lm(reg, x, interval="none")

##        1        2        3
## 5.446950 7.004360 8.561771
```

Scatterplot and Regression Line   To generate the scatterplot, we used the xyplot function in lattice, adding the least squares regression line with the type argument, and a few other aesthetic elements. The regression line has a positive slope, indicating that low physical heath scores are associated with low mental health scores, middle with middle, and high with high, respectively. Notice too that there is a lot of variability in mental health scores *generally*. At every level of physical health, the mental health observations spread out quite substantially around the regression line, which will show up—in specific quantitative terms—in both the RSE and R^2 statistics.

Regression Coefficient β_1 is the slope of the regression line and also summarizes the strength and direction of the *X-Y* relationship. Interpreting $\beta_1 = 0.155$, we can say that self-reported poor mental health increase an average of 0.16 days per month for every 1 additional day of poor physical health. Notice that the interpretability of the regression coefficient depends substantially on the measurement units of *X* and *Y*. If measurement units are real as they are in this example (e.g., days per month), the size of the regression coefficient is easier to understand and evaluate. If the measurement units are points on an arbitrary scale (e.g., 0–10 rating scale), or a scale with which we have no experience, the size of regression coefficient is more challenging to evaluate.

RSE   The interpretability of measurement units applies to evaluating the RSE too. In our example RSE = 9.04, which tells us the average size of the residuals scores, in *Y* units. Think about it this way, if we wanted to predict the mental health (in days per month of poor mental health) of females who smoke based on their physical health scores, our predictions would have an average error of about 9 days per month included in the prediction. Given that most of the mental health observations range between 0 and 20 days (see scatterplot), 9 days seems like a lot of error.

R^2 The coefficient of determination, $R^2 = 0.0.23$ (or 2.3%), is the proportion (or percentage) of the total variance in mental health scores that is accounted for by knowing the participants' physical health. It is also the proportion (or percentage) that the total variance in Y is reduced by using X to predict Y. R^2 is the ratio of explained variance in Y by X to total Y variance (see Formula 5.3). As such R^2 can range from 0.00 (X explains none of the Y variance) to 1.00 (X explains all the Y variance). The larger R^2 is, the better the *model*—which in this example is a measure of physical health—fits the data. As a statistic without real measurement units, R^2 can be challenging to evaluate (i.e., deciding how big of a relationship is indicated by $R^2 = 0.023$).

Predicted Values The regression equation can be used to generate a predicted Y score from any X value. Prediction is a practical, valuable, and widely used application of regression analysis. There are many instances in which professionals need to predict an outcome before it occurs. For example, if educators can predict what learning difficulties students are likely to have next year, from this year's data, they can develop interventions to address (or prevent) those difficulties. We make predictions by substituting values of X into the equation, $\hat{y} = 5.45 + 0.16x$. The predicted values above are the predicted days per month of poor mental health for the three physical health scores entered into the data frame statement. Here is where RSE is very important; we want our predictions to be as accurate as possible because, often, we make decisions based on them. We will learn later in this book how the RSE affects inferential tasks like that.

5.5   Correlation

Regression analysis is the primary, but not the only, statistical method for analyzing data from studies with numeric X and Y variables. Correlational analysis is another common approach. Whereas regression uses a linear function to describe an X-Y relationship, correlational analysis is based on a statistic called the covariance. The covariance of X and Y is simply a measure of how much two numeric variables co-vary, or how much change in X is associated with change in Y. Unlike regression, correlational analysis:

• Does not make a distinction between predictor and outcome variables.
• Does not use a linear function to understand an X-Y relationship.
• Is not used for prediction.

The main appeal of correlational research is that it relies on an intuitive and simple statistic—the correlation coefficient. Like a regression coefficient, the sign of a correlation coefficient conveys the direction, and the magnitude conveys the size, of the relationship. However, the correlation coefficient is scaled so that it ranges from −1.0 to +1.0, with zero indicating no correlation between X and Y. So, the correlation coefficient's strength—and weakness—is that it has no units. A standardized statistic like the correlation coefficient is valuable for comparing findings across studies, when measurement instruments and samples vary. The price we pay for that flexibility, as we saw with R^2 earlier (a similarly-scaled statistic), is the difficulty in interpreting the size of a correlation coefficient.

Let's look at three prominent correlation coefficients and see how they conceptualize covariation.

Pearson r Pearson's correlation coefficient, or Pearson r, as can be seen in Formulas 5.4 and 5.5, is the standardized covariance between numeric X and Y variables. The covariance is merely a special case of the sample variance (see Chap. 1) and multiplies an observation's deviations from the X and Y means, respectively. The covariance is the average of those cross-products for all observations. By dividing the covariance by the product of the X and Y standard deviations, Pearson r converts the unit-specific covariance to a unit-less statistic that is conveniently bounded by ±1.0, with r = 0.00 indicating no relationship (no covariation) between X and Y.

$$cov_{xy} = \frac{\sum (x_i - \bar{x})(y_i - \bar{y})}{N} \tag{5.4}$$

$$r_{xy} = \frac{cov_{xy}}{s_x s_y} \tag{5.5}$$

Spearman Rho Spearman rho is very similar in concept and calculation to Pearson r but takes ordinal (i.e., ranked) X and Y data. The variables can be ordinal to begin with (e.g., high school class rank, Rotten Tomatoes score) or can be transformed from numeric to ordinal. Spearman's rho is simply the standardized covariance of *ordinal* X and Y variables, calculated with the same formulas as Pearson r above. Spearman rho also ranges from ±1.0 with 0 representing no relationship between X and Y. As explained below, Spearman rho is a more resistant statistic than Pearson r.

Kendall Tau Kendall tau is based on the concordance of numeric X and Y variables. What is concordance? Consider two observations with scores (x_1, y_1) and (x_2, y_2). Those observations are concordant if $(x_1 > y_1)$ *and* $(x_2 > y_2)$, or if $(x_1 < y_1)$ *and* $(x_2 < y_2)$. Kendall tau is calculated by finding the total number of concordant and discordant pairs from every pair of observations in the sample data. Kendall tau is simply the difference between the concordant (N_C) and discordant (N_D) pairs divided by the total number of pairs of scores in the sample.

$$\tau = \frac{N_C - N_D}{N(N-1)/2} \tag{5.6}$$

You should be able to see that if all the pairs are concordant, $\tau = 1.0$, if all the pairs are discordant, $\tau = -1.0$, and with equal numbers of concordant and discordant pairs, $\tau = 0$.

Let's generate these correlation coefficients in R and work through the interpretation and implications of each. The cor() function takes X and Y numeric variables, and the method argument specifies which coefficient you want to calculate. Method = "p" will treat the variables as numeric and produce Pearson r, method = "s" will convert those variables to ordinal data and produce Spearman rho, and method = "k" will find concordance/disconcordance in the pairs of X, Y observations and produce Kendall tau.

```
#Pearson r
cor(dat$DaysMentHlthBad,dat$DaysPhysHlthBad,method="p",use="complete.obs")

## [1] 0.1527004

#Spearman rho

cor(dat$DaysMentHlthBad,dat$DaysPhysHlthBad,method="s",use="complete.obs")

## [1] 0.1405031

#Kendall tau
cor(dat$DaysMentHlthBad,dat$DaysPhysHlthBad,method="k",use="complete.obs")

## [1] 0.1150644
```

All three correlation coefficients show a positive, but small, relationship between poor physical health and poor mental health, where increases in one are associated with increases in the other. Again, without meaningful units and without knowing about similar relationships in this research area, it can be hard to evaluate what the size of these correlations mean. The conventions in Fig. 5.4 present some interpretive guidelines that have broad consensus, but keep in mind that even these will vary by discipline. So we can summarize the correlation between poor physical and poor mental health as positive and very weak.

We wrap this section up with two important caveats about correlational methods.

1. Pearson r and least squares regression are non-resistant statistical methods because, statistically, both are built from the mean and variance, which are themselves non-resistant statistics. We know that extreme scores affect the mean and variance and, by extension, Pearson r and the regression coefficient. Fortunately,

Fig. 5.4 Guidelines for interpreting correlation coefficients

correlation	relationship strength
0 - ±.20	none to very weak
±.20 - ±.40	weak
±.40 - ±.60	moderate
±.60 - ±.80	strong
≥ ±.80	very strong

both Spearman rho and Kendall tau are more resistant correlation coefficients. Spearman rho uses ordinal data in its calculations, which moderates the influence of very large or very small values on the coefficient. And Kendall tau is a concordance-based statistic, so extreme values don't really affect the calculations. Because least squares regression is vulnerable to the influence of extreme scores, it is essential that we examine our bivariate data for potential outliers when we use least squares regression. We will see how to do that in the very next section.

2. Pearson r (again, like least squares regression) summarizes only the *linear* element of an *X-Y* relationship. Numeric variables may be related in nonlinear ways, however. Looking again at the scatterplot from our data analytic example above, physical and mental health may well have a nonlinear relationship in addition to the weak positive linear relationship we have already documented. Although this book does not cover nonlinear regression methods, but both Spearman rho and Kendall tau are more sensitive than Pearson r to nonlinear but monotonic (i.e., increasing or decreasing trends) components of an *X-Y* relationship. As a result, they are more accurate correlation coefficients than Pearson r when you suspect the true relationship may be nonlinear.

5.6   Influential Observations

We diagnose influential observations in bivariate data with much the same methods as we did in univariate numeric data, except that we're analyzing residual scores. Remember, residual scores are scores that are "left over" or unexplained after using X to explain, or model, Y. We expect residual scores to vary, but extremely large residuals indicate cases where X explains very little of the total Y deviation score. As a result, observations with large residuals affect overall model fit. Large residuals therefore draw our attention for the same reasons they did in univariate data. First, they have disproportionate influence on regression statistics, including β_1, RSE, and R^2. Second, data points with large residuals suggest the possibility that the observation is from a participant who is not part of the population of interest.

Below is the code we used in Chap. 1 to explore influential observation in a numeric variable using the boxplot rule. The output below shows how to plot the residuals, using the resid() function to extract the residual scores from our regression analysis.

```
#boxplot rule
b=boxplot(resid(reg),horizontal=T,main="residuals from regression analysis")
```

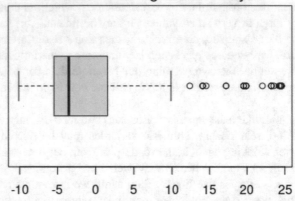

b

```
## $stats
##              [,1]
## [1,] -10.119181
## [2,]  -5.446950
## [3,]  -3.492044
## [4,]   1.553050
## [5,]   9.880819
##
## $n
## [1] 234
##
## $conf
##            [,1]
## [1,] -4.215059
## [2,] -2.769029
##
## $out
##      5037     5434     5594     6036     6108     6207     6405     6406
## 14.55305 19.55305 24.55305 23.55305 24.55305 19.88082 22.06119 17.21693
##      6407     6830     6919     7129     7130     7131     7576     7577
## 17.21693 24.55305 24.24157 23.24157 23.24157 23.24157 24.55305 24.55305
##      8159     8498     9023     9086     9087     9088     9169     9170
## 19.55305 19.55305 24.55305 14.08583 14.08583 14.08583 19.88082 19.88082
##      9202     9342     9343     9344     9716     9730
## 19.55305 24.39731 24.39731 24.39731 19.88082 12.37268
##
## $group
##  [1] 1 1 1 1 1 1 1 1 1 1 1 1 1 1 1 1 1 1 1 1 1 1 1 1 1 1 1 1 1 1 1 1 1 1 1 1
##
## $names
## [1] ""
```

The outlier residual values and their respective case numbers are retrieved from the boxplot output as we did in Chap. 1. As we see above, there were 30 residual scores that exceeded the upper adjacent value of the boxplot 5-number summary. The scatterplot generated at the start of our analysis suggested that there would be some observations with large residuals, due mainly to the large variance in mental health scores.

5.7 Plots

The scatterplot is the primary graphical tool for exploring and summarizing bivariate numeric data. In our data analytic example above, we used the scatterplot function xyplot() from the lattice package. Here I demonstrate a different function, the scatterplot() function from the car package. The scatterplot() function plots univariate boxplots on the X and Y axes for each variable, which adds to the plot's utility for the analyst as well as for communicating the results of regression or correlational analysis. The example below shows the toggles to control the appearance of the regression line in the regLine argument, as well as axis and plot labels.

```
library(car)

library(car)

## Loading required package: carData

## Loading required package: carData
scatterplot(DaysMentHlthBad ~ DaysPhysHlthBad, data=dat,
            smooth=F,
            regLine=list(method=lm, lty=1, lwd=1, col="red"),
            xlab="Self-reported poor physical health (days)",
            ylab="Self-reported poor mental health (days)",
            main="Relationship between poor physical and
         mental health in females who smoke (NHANES data)")
```

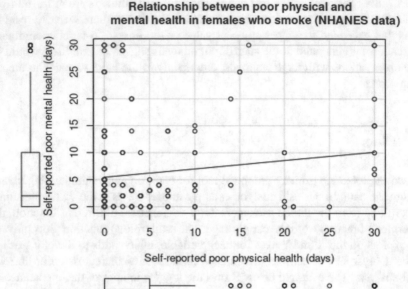

The boxplots show that univariate outliers (by the boxplot rule) do not necessarily translate into outlying residual scores from a least squares regression analysis. We also notice that people tend to use round numbers when self-reporting quantities like days per month of poor physical health.

5.8 Interpreting Correlations

"Correlation does not imply causation" is an oft-repeated caveat about interpreting correlational research, although its admonition is commonly disregarded even by experienced researchers when interpreting findings from correlational research. To understand what the caveat means, let's consider the causal possibilities that exist if two variables are correlated. Say we're interested in the effect of stress (X) on health (Y), and we measured those two variables in an observational study. We do a correlational analysis of the data and find Pearson r = −0.50. We know that is evidence of covariation, but does that correlation provide any evidence of the causal *effect* of stress on health?

There are three causal possibilities that could produce correlation between two variables:

1. $X \longrightarrow Y$

2. $Y \longrightarrow X$

3. $X \longleftrightarrow Y$
        $\swarrow \nwarrow$
          Z

In terms of our example, #1 shows the causal effect of stress on health, #2 shows the causal effect of health on stress—which is equally plausible and likely—and #3 represents all the "third variables" that could be associated with both greater stress and poorer health, such as personality or lifestyle variables. Repeating principles we have learned about interpreting statistics in previous chapters, it's the design of the study that affords causal interpretations, not a particular statistic. The higher the study's internal validity (see Chap. 3), the more confidence we have that whatever relationship we observe—in our example above a weak positive relationship—is due to the predictor variable and the predictor variable alone. In our example looking at the relationship between physical and mental health in female smokers, internal validity is very low because (a) the predictor variable was measured and not experimentally arranged and (b) none of the alternative explanations (like the possibilities in #2 and 3 above) have been controlled.

When variables have temporal order—such as when X occurs before Y—the design of the study rules out causal possibility #2. If, for example, we measured stress in September and health in December, it is not possible for the later event to cause the earlier one. This longitudinal design feature improves the study somewhat, but it still has not controlled for the many predictors of both stress and health that occur before September. Longitudinal studies that measure the predictor variable at time 1, the outcome variable at time 2, *and* control for plausible common causal variables (i.e., alternative explanations) are strong designs and afford the researcher some basis for making causal conclusions with correlational statistics.

5.9   Data Analytic Example 2

For our second data analytic example, we work with the environmental dataset from the lattice package, which includes measures of atmospheric conditions in New York City from 1973 (see the documentation for variables and measurement details). Let's investigate the relationship between daily high temperature (X, in degrees Fahrenheit) and average daily ozone concentration (Y, in parts per billion). This example also demonstrates a data analytic procedure not covered in example 1: how to identify boxplot-defined outliers in a regression model by case number and then exclude them from regression analysis.

The chunk of output below shows the following, in order:

- A scatterplot of the temperature-ozone relationship with the least squares regression line plotted on it.
- The regression coefficients, RSE, R^2, and the Pearson and Spearman correlation coefficients for the relationship between temperature and ozone.
- A boxplot of the residuals from the regression model, with the outliers labeled by case number. Note that we use the Boxplot() function from the car package with an argument in the function that assigns case (or row) numbers to all outliers.
- The regression coefficients, RSE, R^2, for the adjusted model (outliers removed from the analysis). We do this with the subset argument in the lm() function, with the –c() function meaning "remove the following cases."

```
#scatterplot w/regression line
library(survival)
library(lattice)
xyplot(ozone ~ temperature, data=environmental,
       grid=TRUE,
       type=c("p", "r"),      #p=scatterplot, r=regression line
       col.line="red")
```

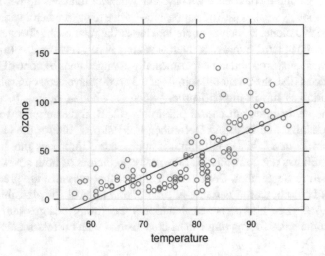

```
reg=lm(ozone ~ temperature, data=environmental)
reg$coefficients
```

```
## (Intercept) temperature
## -147.64607      2.43911
```

```
summary(reg)$sigma
```

```
## [1] 23.92025
```

```
summary(reg)$r.squared
```

```
## [1] 0.4879601
```

```
cor(environmental$ozone,environmental$temperature,method="p",use="complete.obs")
```

```
## [1] 0.6985414
```

```
cor(environmental$ozone,environmental$temperature,method="s",use="complete.obs")
```

```
## [1] 0.7729319
```

```
#find outlier residuals and their case numbers
library(car)
Boxplot(resid(reg), id.method="y")
```

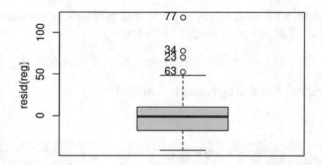

```
## [1] 23 34 63 77

reg2=lm(ozone ~ temperature, data=environmental, subset = -c(23,34,63,77))
reg2$coefficients

## (Intercept) temperature
## -137.967521    2.276021

summary(reg2)$sigma

## [1] 17.86539

summary(reg2)$r.squared

## [1] 0.602601
```

What do we learn from this analysis? The regression coefficient tells us that for each additional degree in daily high temperature, ozone concentrations increase by an average of 2.44 parts per billion. The two model fit statistics tell us that daily high temperature explains about 49% of the total variance in ozone concentrations and that the average error of prediction (if we were to use the model to predict ozone concentration from temperature) would be 23.9 parts per billion. The correlation coefficients help communicate the size of the relationship, inasmuch as they are scaled to a 0–1.0 scale. Both correlations indicate a moderate to strong relationship between daily high temperature and ozone concentrations, but the Spearman coefficient (rho = 0.77), which is the more resistant correlation coefficient, indicates a stronger relationship than the Pearson coefficient (r = 0.70). Why? Remember, the Spearman coefficient is more sensitive to nonlinear elements of an X-Y relationship, and the scatterplot of the residuals suggests that a curve might actually fit the data better.

However, we need to determine what, if any, influence outlying residual scores have on the model fit. The boxplot of residual scores from the regression analysis reveals four outlier values, with their respective case numbers. Those outliers show very high ozone readings on days when the temperature is not extremely warm, suggesting that some other factor(s) may have affected the ozone levels on those days. When we remove those four data points and redo the analysis, the regression coefficient ($\beta_1 = 2.28$) shows a slightly smaller average change in ozone for each degree in daily high temperature. But the fit statistics (RSE = 17.9, $R^2 = 0.60$) reveal a much better overall fit, with substantially larger R^2 and smaller RSE statistics. A written summary of the data analytic example 2 is below.

5.10   Writing Up a Regression Analysis

Results

The relationship between daily high temperature (°F) and ozone concentration (ppb) in New York City was examined in data from May to September of 1973 (N = 111). A scatterplot of the scores is in Box Fig. 5.1. The least squares regression coefficient ($\beta = 2.44$) showed ozone concentration increased about 2.4 ppb for each degree increase in daily high temperature. Daily high temperature explained about 49% of the variability in ozone concentration ($R^2 = 0.49$). Analysis of the residuals revealed a substantial amount of unexplained variance (RSE = 23.9) as well as four observations with unusually large residuals. When those scores were removed from the analysis, the relationship was largely unchanged ($\beta = 2.28$), but model fit improved substantially ($R^2 = 0.60$, RSE = 17.9).

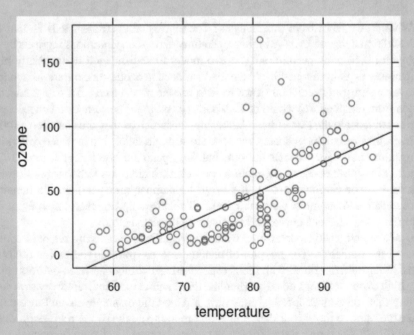

Box Fig. 5.1 Daily high temperature and ozone concentration in New York City (May through September 1973)

5.11   Problems

The problems below all use datasets in the MASS package. Use `library(MASS)` *to make the datasets available and*? `datasetname` *to see the documentation for each, which will provide variable names and measurement details.*

1. In the `cats` dataset, generate a scatterplot of the `Bwt` (x) and the `Hwt` (y) data, with the least squares regression line. Do the least squares regression and interpret β, RSE, and R^2.

2. In the problem above, analyze the residuals for influential observations using the boxplot rule. What case(s) did you identify? What are their values on `Bwt` and `Hwt`? Offer a possible explanation for that observation's large residual.

3. In the problem above, find and interpret Pearson r, Spearman rho, and Kendall tau. What do they summarize about the relationship between body weight and heart weight?

4. In the `UScereal` dataset, generate a scatterplot of the `carbo` (x) and the `calories` (y) data, with the least squares regression line. Do the least squares regression and interpret β and RSE. Now, use the subset argument in the lm() function to drop the two cereals with unusually high carbohydrate content from the analysis, like this:

```
lm(calories~carbo, data=UScereal, subset=UScereal$carbo<40)
```

Does the exclusion of those two observations change β and RSE? Explain why not.

5. In the `UScereal` dataset, generate a scatterplot of the `sugars` (x) and the `calories` (y) data, with the least squares regression line. Interpret β and RSE. Notice the two cereals with extremely high calorie levels (about 360 and 440). Confirm that they are influential data points with a boxplot test of the residuals. Remove those two observations from the analysis using the argument below in your lm() model statement.

```
subset=UScereal$calories<300
```

Does the exclusion of those two observations change β and RSE? Explain why.

6. In the `Boston` dataset, generate a scatterplot of the `rms` (x) and the `medv` (y) data. Find Pearson r and Kendall tau for the relationship between rooms and median value of house. What do they tell you? What might be the reason for the difference between the two correlation coefficients?

7. In the #6 above, do a least squares regression using rooms to predict median value. Interpret β, RSE, and R^2 for your regression model. Use the regression

equation to predict median value for a house with five rooms and eight rooms, respectively.

8. Is there a relationship between mother's weight (x) and infant birth weight (y)? Using the data in the birthwt data, do a regression analysis; interpret β, RSE, and R^2. Also explore the residuals for influential values. Generate a scatterplot with the least squares regression line, and summarize your findings in a written paragraph.

Chapter 6
Statistics and Data Analysis in a Logistic Model

Covered in This Chapter
- Statistics and plots for logistic model data
- Maximum likelihood estimation and simple logistic regression
- Common elements between least squares and logistic regression

6.1 Data Analysis in a Logistic Model

In this chapter we learn how to deal with data from studies with a numeric predictor variable and a categorical outcome variable, or a logistic model. Recall, a logistic model consists of numeric X and a categorical Y variable.

$$\text{categorical } Y = \beta_0 + \beta_1 \text{numeric } X + \varepsilon$$

In this chapter we will deal with ANOVA models in which X is a two-category variable. Many research questions take this form, such as:

- Does number of AP credits earned in high school affect being admitted to college (admit/not admit)?
- Does credit card balance predict making a late payment (late/on time) on one's credit card bill?
- Does body weight predict being diagnosed with hypertension (yes/no)?

In Chap. 5 we used regression methods—least squares regression, to be precise—to describe the relationship between numeric x and y variables. We will use regression analysis again to address the research questions above but employ a different method—*logistic regression*—to deal with the statistical peculiarities of two-category outcome variable. With least squares regression, we use a linear function to *predict*: the regression equation generates an average predicted value on the

© The Author(s), under exclusive license to Springer Nature
Switzerland AG 2023
B. Blaine, *Introductory Applied Statistics*,
https://doi.org/10.1007/978-3-031-27741-2_6

(numeric) outcome variable from predictor variable information. With logistic regression we use a linear function to *classify*: the regression equation generates a predicted probability of being in the target category from predictor variable information.

That key difference aside, many of the methods and concepts in this chapter carry over from Chaps. 4 and 5. Chapter 4 introduced statistics like the proportion, odds, and odds ratio. Chapter 5 introduced statistics like the regression coefficients, RSE, and R^2 as important and useful in quantifying that relationship. We use all of these concepts again in this chapter, with a few tweaks here and there.

6.2 Logic of Logistic Regression

To understand how logistic regression works to summarize a relationship between a numeric X and a categorical Y, it helps to first see how least squares regression doesn't work very well with categorical outcome data. Let's imagine we want to predict the sex of a cat from its heart weight (in kgs). Using the cats dataset (MASS package), let's look at a scatterplot of heart weight (X) and sex (Y, 0 = female, 1 = male) with the least squares regression line (Fig. 6.1). As we did in Chap. 4, consider the data points as probabilities of being in category 1 (being a male cat). We can see from the data that heavier hearts tend to be from male cats, which is to say that as heart weight increases the probability of it being from a male, compared to a female, cat increases. This general relationship is reflected in the positive slope of the least squares regression line, which by the way is fit to the data exactly as we learned in Chap. 4—by minimizing the sum of squared residual scores.

However, there are two problems with using a linear function to summarize a relationship with a categorical outcome variable.

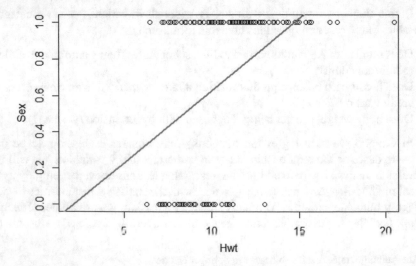

Fig. 6.1 Least squares regression line fit to heart weight (X) and sex (Y) data

- Problem 1: Notice that linear regression can generate predicted probabilities of the outcome that are outside of its 0 to 1 range. The regression line extends beyond the plot in both directions, and you can see that heart weights in excess of about 16 kgs lead to predicted probabilities greater than 1. This doesn't make sense because 1 is merely a code for the male category; there are no categorical "values" outside of 0 and 1. Similarly, the probability of *any* outcome can only exist between 0 and 1, so predicted probabilities outside of that range are absurd.
- Problem 2: Notice that a linear function establishes a constant amount of change in Y (the probability of being a male cat) for a 1-unit increase in X. In this example that would mean that for each 1 kg increase in heart weight, the probability of being a male cat increases by some constant amount wherever you are on the heart weight scale. However, the true relationship between heart weight and probability of being a male is such that the amount of change in Y varies, depending on the particular value of X. At low levels of heart weight (say between 6 and 9 kg), there are many more female than male cats, so a 1-kg increase doesn't change the probability of that case being a female cat much. Similarly, almost all the cats with heart weight greater than 12 kg are male, so a 1-kg increase would not change the probability of those cases being a male cat much, if at all. In the middle heart weight values (say 9–12 kg), however, the probabilities of being a male change more with a 1-kg increase in weight.

To better approximate, or *model*, the true relationship between heart weight and sex, we need a nonlinear function that does two things: first, it must not make predictions greater than 1 or less than 0, and second, it must capture the changing influence a 1-unit increase in X has on the probability of Y, depending on where that change in X is on the X scale. There are many nonlinear functions that could meet these criteria, but the logistic function is a computationally simple and interpretable function that works well for data of this type (numeric X, categorical Y). Figure 6.2 shows the logistic function fit to the heart weight and sex data.

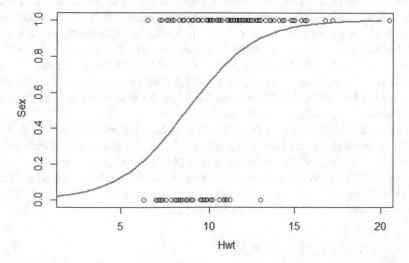

Fig. 6.2 Logistic regression line fit to heart weight (X) and sex (Y) data

Table 6.1 Probabilities transformed to log odds

p	0.1	0.2	0.3	0.4	0.5	0.6	0.7	0.8	0.9
1-p	0.9	0.8	0.7	0.6	0.5	0.4	0.3	0.2	0.1
Odds	0.11	0.25	0.43	0.67	1	1.5	2.33	4	9
log odds	−2.2	−1.4	−0.9	−0.4	0	0.41	0.85	1.4	2.2

Source: Pampel, F. (2000). *Logistic regression: A primer.* Quantitative Applications in the Social Sciences, Vol 132. Sage Publications

The basic logic of logistic regression is this: Instead of trying to fit a logistic function to the data (which would involve estimating several parameters because it is a cubic function), we will log-transform our outcome variable in such a way that the transformed outcome variable is a linear function of X (which then would involve estimating just one parameter—the slope) and then find the best-fit line. Below we walk through the steps of this logic in statistical terms.

In logistic regression we first transform our outcome variable using the **logit** (L) function.

$$L_i = \ln\left(p_i / 1 - p_i\right)$$

(6.1)

The logit function first takes the probability of the outcome (p), then finds the odds of the outcome ($p/1-p$), and then finds the log of the odds of the outcome. The transformed statistic, L_i, is called the *log odds* or the **logit**. Let's look at a table of transformed values (from p to log odds) to see what the logit function does for us (Table 6.1).

First, recall from Chap. 3 that an event with p = 0.5 has an odds of 1. Converting probabilities to odds eliminates the upper limit that probabilities impose on our data (i.e., probabilities can't be greater than 1.0). The lower limit (i.e., probabilities can't be less than 0) is eliminated by converting odds to log odds. So, notice that the log odds scale is symmetrical around 0 (indicating even odds, where the event is equally likely to occur or not occur) and extends to infinity in both directions. The logit function meets the first requirement of a function that doesn't make predictions greater than 1 or less than zero. As p gets large, so does the logit, but since the log odds of p = 1 is undefined, the logit approaches but never reaches p = 1. The same thing occurs for low probability events; the log odds of p = 0 is undefined, so the logit function approaches but never reaches p = 0.

Second, it is very important to realize that the logit transformation linearizes the nonlinear relationship between a numeric predictor and a categorical outcome. It does this by "stretching" the probabilities of Y (when Y is near 0 or 1) so that a 1-unit increase in X is associated with the same amount of change in the log odds of Y at any point along the X axis. This property of the logit addresses the second requirement for a function that models the decreasing influence a 1-unit increase in X has

Table 6.2 Log odds (L) transformed to probabilities

log odds	−3	−2	−1	0	1	2	3
p	0.047	0.119	0.269	0.5	0.731	0.881	0.953

Source: Pampel, F. (2000). *Logistic regression: A primer*. Quantitative Applications in the Social Sciences, Vol 132. Sage Publications

on the probability of Y, at low and high values of p. The linearizing of a nonlinear relationship is important for a second reason: a nonlinear relationship cannot be summarized with a single coefficient, but a linear relationship can.

We can see both of these properties of the logit by looking at a range of logit values that are transformed back to probabilities (see Table 6.2). First, larger log odds values in both directions approach p = 0 and p = 1, respectively, but never actually reach those values. Second, changes in the probability of the outcome depend on where you are on the log odds scale. Going from L = 0 to L = 1 is associated with a larger change in the probability than going from L = 2 to L = 3. This demonstrates the nonlinear relationship between L and p.

Let's recap these principles:

- A numeric predictor has an inherently nonlinear relationship with a categorical outcome, so a linear function results in a poor fit.
- The logit function preserves the two essential requirements of that nonlinear relationship: avoiding predicted values beyond the range of possible outcomes and capturing the variable influence of changes in X on the log odds at extreme values of p.
- The logit function also linearizes that relationship and allows us to describe the relationship between X and log-odds(Y) with a linear function and a single coefficient (i.e., slope of the regression line), just as we do with least squares regression.

With that established, next we need to learn how a linear function is fit to the X-log odds(Y) data and what statistics we use to describe relationships in logistic regression.

6.3   Logistic Regression Analysis

In this section we learn the estimation method and statistics inherent to logistic regression analysis. There are clear parallels in the statistical concepts underlying least squares and logistic regression (see Table 5.3), so your work in Chap. 4 will pay off here. Although the statistical concepts are calculated differently in logistic regression because of the logit transformation discussed earlier, we will draw those parallels so that you can elaborate your knowledge of regression rather than try to learn a "new" procedure from scratch.

Maximum Likelihood Estimation

Recall that in least squares regression, the optimal fit had to do with finding the linear equation (with its particular slope and intercept) that minimized the sum of squared residual scores. In logistic regression the optimal fit has to do with finding the linear equation that generates predicted probabilities that are closest to the actual probabilities. The method for finding the optimal regression line is called maximum likelihood estimation. The concepts of least squares and maximum likelihood are essentially the same, but the computations in maximum likelihood are not as simple, so let's work through that next.

$$LF = \prod \left\{ p^y * \left(1 - p^{1-y}\right) \right\}$$

(6.2)

Maximum likelihood estimation is driven by the likelihood function (Formula 6.2), in which Y = the actual probability of a case (so, either 0 or 1) on the outcome variable for a particular X value and p = the predicted probability for that X value generated by the logistic regression line. The term inside the curly braces is the *likelihood score* for a particular case. Notice the parallels with residual scores in least squares regression, where we also capture the difference between actual Y and predicted Y.

The upper case pi symbol ($\prod$) is a sign for continued multiplication, indicating that all of the likelihood scores will be multiplied. Maximum likelihood estimation finds the linear equation (with particular slope and intercept) that *maximizes* the likelihood function. Look back at Fig. 6.2: you can see that, for a given x value, the closer the predicted value is to the actual value, the *greater* the product of those probabilities will be. This is the parallel idea to the least squares regression line that *minimizes* the sum of squared residuals (difference between predicted and actual Y scores).

Table 6.3 Actual and predicted probabilities from good-fitting and poor-fitting models and their implications for likelihood and log-likelihood scores

y_i	p_i	$p_i^{y_i}$	$(1-p_i)^{1-y_i}$	L	LL
Good-fitting model					
1	0.8	$0.8^1 = 0.8$	$0.2^0 = 1$	0.8	−0.223
1	0.7	$0.7^1 = 0.7$	$0.3^0 = 1$	0.7	−0.357
0	0.3	$0.3^0 = 1$	$0.7^1 = 0.7$	0.7	−0.357
0	0.2	$0.2^0 = 1$	$0.8^1 = 0.8$	0.8	−0.223
Poor-fitting model					
1	0.2	$0.2^1 = 0.2$	$0.8^0 = 1$	0.2	−1.61
1	0.3	$0.3^1 = 0.3$	$0.7^0 = 1$	0.3	−1.20
0	0.7	$0.7^0 = 1$	$0.3^1 = 0.3$	0.3	−1.20
0	0.8	$0.8^0 = 1$	$0.2^1 = 0.2$	0.2	−1.61

L likelihood, LL log-likelihood

Source: Pampel, F. (2000). *Logistic regression: A primer*. Quantitative Applications in the Social Sciences, Vol 132. Sage Publications

Let's look at how the likelihood function works in two scenarios. A score's likelihood is the probability that a regression model (slope and intercept) could generate the actual score from its X value. Table 6.3 shows 4 scores and their likelihoods from a good-fitting regression model and the same scores with likelihoods generated by a poor-fitting model. Notice that if we multiplied these likelihood scores (in the L column), we would get a larger number in the good-fitting model scenario than the poor-fitting model. So, the best-fitting model is the one (with particular slope and intercept) that results in the maximum likelihood, as generated by the ML function above.

The principle of maximum likelihood is perfectly fine for finding the "best fit" line in a logistic regression analysis. But some of our model statistics (see deviance and R squared below) emerge from *minimizing* the *sum* of the log-likelihood scores rather than *maximizing* the *product* of the likelihood scores. Formula 6.3 shows the log-likelihood function, which has simply taken the log of both sides of the maximum likelihood function above and adds instead of multiplies the terms.

$$\ln LF = \Sigma\left\{\left[y * \ln\left(p\right)\right] + \left[\left(1 - y\right) * \ln\left(1 - p\right)\right]\right\}$$

(6.3)

Why do this? The log-likelihood (or, LL) avoids the very small numbers—and the large numbers of decimal places (and possible rounding errors)—that result from multiplying likelihood values in the maximum likelihood function. Whereas the likelihood ranges from 0 to 1, and can be a very small number, the LL ranges from zero to negative infinity. If the model perfectly reproduces the actual scores with the predicted scores, $LL = 0$. So the smaller LL, the better. LL will increase in the negative direction as the model generates less accurate estimates of the true values, or in other words as model fit decreases. Looking at the example in Table 6.3, the better fitting model results in smaller LL scores, which of course sum to a smaller number. The log-likelihood is the basis of several important statistics in logistic regression analysis. Let's look at them.

Deviance The deviance is the equivalent statistic in logistic regression to the RSE in least squares regression. The deviance captures the fit of the model to the data. In other words, how close is the model's predictions for a case's outcome to its actual outcome? The more accurate the predictions, the smaller the deviance. The deviance statistic multiplies the LL by -2 and is sometimes symbolized as "-2LL." The reasons for that rescaling don't concern us in this book. Just know that the deviance is a descriptive statistic representing the overall adequacy of the model to reproduce the actual scores. While the units of the deviance are not intuitive and not easy to interpret, that won't be a problem because we will be comparing deviances from models to see how much they change.

R^2_{logistic}

The logistic regression equivalent of the R^2 statistic in least squares regression finds the proportion of the total deviance in the intercept-only model that is reduced by using the predictor variable. Remember in least squares regression when you didn't have any predictor to work with the best way to predict someone's Y value was to use the mean of Y? It's the same idea here: with no predictor, the intercept (the Y mean when $X = 0$) makes the best prediction. How much that intercept model deviance is reduced by using X to predict Y is the essence of R^2 in logistic regression. As with the coefficient of determination in least squares regression, $R^2_{logistic}$ can range from 0 to 1.0 (or 0% to 100%).

Logistic Regression Coefficient

We learned earlier that although our raw data are probabilities (0s and 1s), we convert those to odds and then to log odds for analysis. Accordingly, logistic regression coefficients (i.e., the slope and intercept of the best fit line as defined by the ML function) are in log odds—and interpreted as the average change in the log odds of Y for a 1-unit increase in X. These are not very user-friendly statistics to interpret, so to render the regression coefficients in more interpretable terms than the log odds, we convert them to odd ratios. The odds ratio (OR) is familiar territory. As we learned in Chap. 3, the odds ratio tells us how much the odds of being in the 1 compared to the 0 category change with a 1-unit increase in the predictor variable.

Let's recap the main points of this section:

- Maximum likelihood estimation follows the maximum likelihood function and finds the linear equation that generates predicted probabilities that are closest to the actual probabilities. Maximum likelihood estimation is equivalent to least squares estimation.
- Log-likelihood scores eliminate the very small numbers inherent to maximum likelihood estimation and provide the basis for the deviance and R squared statistics.
- Deviance quantifies the overall ability of a logistic model to reproduce the actual scores, and the smaller the better. Deviance is an equivalent statistic to RSE in least squares regression.
- The reduction in deviance from using the intercept to using the predictor to predict actual scores forms the basis of R squared in logistic regression.
- Logistic regression coefficients are analyzed in log odds but interpreted and reported in odds ratios.

Point-Biserial Correlation Coefficient

In Chap. 5 we saw how data from numeric X and Y variables could be summarized with a correlation coefficient. The main appeal of correlation statistics is that they convey the direction and magnitude of an X-Y relationship on a scale that ranges from -1.0 to $+1.0$, with zero indicating no correlation between X and Y. This standardized scale can be useful for comparing findings across studies, when measurement instruments and samples vary, but its lack of real-world units can make it less interpretable than a regression coefficient. Correlational analysis can also be done with data from a logistic model, provided the outcome variable is a numeric variable (e.g., coded 0, 1). The formulas for calculating the Pearson r can be used to produce a correlation coefficient between a numeric X (with many values) and a numeric Y (with 2 values). Applying Formulas 5.4 and 5.5 to this type of data generates the *point-biserial correlation coefficient*, symbolized r_{PB}, and it is interpreted as any correlation coefficient.

6.4   Data Analytic Example 1

Let's do an example of logistic regression analysis using 2011–2012 NHANES data in R. We will analyze the relationship between days per month of self-reported poor physical health (X) and sleep trouble (Y, yes/no). Our population of interest for this question is married survey participants. As we did in Chap. 4, we must designate X and Y variables for regression analysis and do so with some awareness of their plausible causal connections. In this example poor physical health causing sleep trouble and sleep trouble causing poor physical health are both plausible, as is the possibility of other uncontrolled variables (e.g., underlying illness) being related to both X and Y[1].

The chunk of output below shows the following data analytic operations, in order:

- Address two data wrangling tasks: select the 2011–2012 data and select only the married participants' data for analysis.
- Fit the logistic regression model and retrieve from it the descriptive statistics presented above.
- Generate a plot whose primary purpose is to remind us that logistic regression is *linear* regression and the plot displays the linear relationship between X and Y on a log scale.
- Use the model to predict the probability of sleep troubles (our target or "1" category of the outcome variable) for several values of the predictor.

Let's do the analysis all at once and then work through the interpretations and implications of the results.

---

[1] As a brief aside, notice that the X-Y designation in bivariate research determines what statistics and descriptive methods are needed. If we had designated sleep trouble as X and days per month of self-reported poor physical health as Y, we would be in an ANOVA rather than a logistic model.

```
library(NHANES)
dat<-NHANES[NHANES$SurveyYr=="2011_12",]
#subset rows by levels of a factor
#dat<-dat[which(dat$MaritalStatus=="Married"),]


library(survival)

#do logistic regression and extract statistics from glm object

#y~x formula statement
reg=glm(SleepTrouble ~ DaysPhysHlthBad,data=dat,family=binomial)

exp(reg$coefficients)   #convert log-odds to OR

##      (Intercept) DaysPhysHlthBad
##        0.3011476       1.0515832

reg$null.deviance   #null deviance (intercept-only model)

## [1] 4035.866

reg$deviance   #deviance after using x

## [1] 3929.616

Rsq<-((reg$null.deviance)-(reg$deviance))/reg$null.deviance
Rsq

## [1] 0.02632655

#correlation coefficient
#point biserial r
cor.test(dat$DaysPhysHlthBad,as.numeric(dat$SleepTrouble))$estimate

##        cor
## 0.1847584

#scatterplot w/linear function and log probabilities
library(effects)

## Loading required package: carData

## lattice theme set by effectsTheme()
## See ?effectsTheme for details.

plot(allEffects(reg,se=F))
```

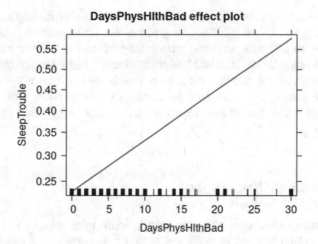

```
#predict from model
predict(reg, data.frame(DaysPhysHlthBad=c(5,10,15)), type="response")

##          1         2         3
## 0.2791522 0.3324362 0.3903829
```

The glm() function finds the linear equation (with particular slope and intercept) by the maximum likelihood function and method we learned earlier. Because the glm() function generates the regression coefficient in log odds, we convert that coefficient back to an odds ratio (using the exp() function) for easier interpretation. What do we learn from these statistics? There is a positive but small X-Y relationship (OR = 1.05). Interpreting the OR, we know that a 1-day increase in days per month of self-reported poor physical health is associated with higher odds of having sleep trouble (because OR < 1). How much higher?—about 5% higher. Or, we could say that the odds of having sleep trouble (compared to none) increase by a factor of 1.05 for every additional day a participant reports having poor health.

The next two commands retrieve the deviance statistics from the model summary. The null deviance is the total "error" or lack of fit in the intercept-only model. The model deviance is the amount that remains after using the predictor (in this example, poor physical health) to predict sleep trouble. The absolute difference between null and model deviance, however, is difficult to evaluate. In later chapters we will learn methods for determining if that change is "significant," but for now we will use the deviance in the form of the logistic R squared statistic, which is generated following the formula learned earlier. $R^2_{logistic} = 0.03$, which tells us that there is only about a 3% reduction in the model deviance when participants' sleep trouble is explained by their poor physical health over and above the intercept-only (or null) deviance. The point-biserial correlation ($r_{PB} = 0.18$) is positive, consistent with an OR > 1, and all three statistics (OR, $R^2_{logistic}$, r_{PB}) indicate a small relationship.

As mentioned above, the plot shows the optimal linear relationship between the probability (plotted on a logarithmic scale) of having sleep trouble and poor physical health. While the plot doesn't communicate model fit, it allows us to see approximate predicted probabilities for given values of the predictor. Our last piece of analysis is to generate predicted probabilities of having sleep trouble for three values of the predictor (5, 10, and 15 days per month). These probabilities show the overall positive *X-Y* relationship (as X increases the probability of Y increases) but also that the probability of sleep trouble doesn't change much with each 5-day increase in poor physical health. This reflects the small effect size captured in the OR and R^2_{logistic}.

6.5   Influential Observations

As with least squares regression, we analyze influential observations in logistic regression through the lens of residual scores. Large residual scores can have disproportionate influence on summary statistics such as the regression coefficient, deviance, and R^2_{logistic}. Recall that in least squares regression a residual score was the difference between predicted and actual *Y* for a given *X* value. In logistic regression, predicted and actual *Y* values are probabilities, and the fact that they can range only between 0 and 1 poses a problem for looking at the distribution of residual scores for values that are too large. We address this problem by generating and retrieving *standardized* residual scores from our logistic model. Scores are standardized when they are converted from their original metric (or, raw scores) to another metric that has properties that are particularly useful or user-friendly. For example, raw IQ test scores are often standardized by converting them to a scale with a mean of 100 and a standard deviation of 10. This way, it is pretty easy to know if an IQ score is above or below the mean and by how much (in standard deviation terms). Similarly, in education raw test scores are commonly converted to a scale with a mean of 50 and a standard deviation of 10; these are called *T*-scores.

To analyze residuals from a logistic regression for influence, we will generate and use *Pearson residuals*. Pearson residuals are raw residuals (i.e., difference between predicted and actual probabilities of *Y* for a given *X*) converted to a scale with mean = 0 and standard deviation = 1. This is a more user-friendly scale because, armed with a basic understanding of the normal curve and its properties, we know that scores that are greater than ±3.0 (i.e., 3 or more SDs above or below the mean) are very unusual and potentially highly influential. Below we generate Pearson residuals from our regression model in the example above and examine them with a boxplot, as we did with residuals in least squares regression. As in Chap. 5, we use the boxplot function from the car package to retrieve the case numbers of any boxplot-defined outliers.

```
#boxplot rule
#find outlier residuals and their case numbers
library(car)
res<-residuals(reg,type="pearson")
Boxplot(res, id.method="y")
```

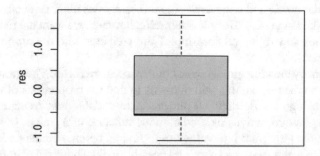

Let me make a brief note in this section about *unbalanced* outcome data in a logistic regression model. In our data analytic example, the distribution of cases across the sleep trouble categories was fairly balanced (see below), with about 73% and 27% reporting no and yes, respectively.

```
table(dat$SleepTrouble)

##
##   No  Yes
## 2824 1019

prop.table(table(dat$SleepTrouble))

##
##        No       Yes
## 0.7348426 0.2651574
```

However, when cases are severely unbalanced in the outcome variable, the maximum likelihood function does not produce accurate estimates of regression coefficients, and the point-biserial correlation coefficient is restricted in the range of values it can take on. A good rule of thumb is that if the distribution of cases is 5%/95% or worse (more unbalanced), an alternate data analytic method should be used and those methods go beyond the scope of this book. Fortunately, extreme unbalance in categorical outcome data is rare.

6.6   Plots

In Chap. 4 we used grouped barplots of proportions to display bivariate data with a categorical outcome variable. Of course, that tool made sense because we also had a categorical predictor. With a numeric predictor, we still want to display the proportions of the events and nonevents (1 and 0 categories on Y, respectively) as they vary across the range of X. Since X is numeric, however, we want the plot to capture the continuous underlying relationship of the predictor with the proportion of the outcome variable.

We do this by creating groups out of our numeric predictor. The group boundaries will be continuous, but this will allow us to find the proportions of the event of interest in each group. By dividing our predictor variable into enough groups, the plot will display proportions as a continuous function of X value. In Chap. 1 we fiddled with the histogram interval width to find an optimal amount of smoothing, creating a plot that was as transparent as possible to the underlying continuous relationship. We follow the same principle here.

To do this we call on a useful function (cut2(), in the Hmisc package) for rendering a continuous variable into groups. This function creates quantile-based groups, resulting in a set of predictor variable "categories" that are continuous across the range of X and have approximately the same number of cases in each. Because our predictor's distribution is dominated by 0s and s, we "cut" our predictor values by tertiles, or into three groups. Then we create a frequency table, with X in columns and Y in rows. And finally we convert that table to proportions for plotting. The plot and the proportions table show a fairly monotonic albeit small negative relationship between self-reported days of poor physical health and the probability of having sleep troubles.

```
library(Hmisc)

#g=3 in cut2() creates tertile-based groups
dat$badhealth=cut2(dat$DaysPhysHlthBad,g=3)
tab=table(dat$SleepTrouble,as.factor(dat$badhealth))
addmargins(tab)

##
##           0    1 [2,30]  Sum
##   No   1803   65    692 2560
##   Yes   499   35    392  926
##   Sum  2302  100   1084 3486

p=prop.table(tab,margin=2)
p

##
##              0         1    [2,30]
##   No  0.7832320 0.6500000 0.6383764
##   Yes 0.2167680 0.3500000 0.3616236

barplot(p,col=c("blue","red"),
        ylim=c(0,1),beside=1,
        legend=rownames(tab),args.legend=list(x="topright"),
        main="Proportion of participants reporting sleep trouble
        by days per month of bad physical health")
```

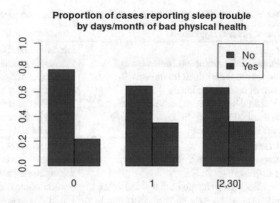

Proportion of cases reporting sleep trouble by days/month of bad physical health

6.7   Interpreting the Results of Logistic Regression

This is the second of two chapters in which we learned how to use regression analysis to analyze bivariate data. We have learned two regression models: often these models are referred to as a *linear model* (implying a numeric Y variable) and a *logistic model* (implying a categorical Y variable). Our study in this chapter helped us see that logistic regression is, in fact, linear in its computations but nonlinear in its essence and purpose. Despite their using different statistics, linear and logistic regression models have common conceptual elements; these commonalities are organized in the table below (Table 6.4).

What are the guidelines for interpreting regression coefficients (e.g., beta, OR) and regression model statistics (e.g., R squared, deviance)? We introduced the important interpretive principles in Chap. 4 when we discussed what causal conclusions can and cannot be drawn from a correlation. These principles—temporal ordering of predictor and outcome variables and control of common causal variables—apply to interpreting regression statistics too. Let's review them.

It is obvious that numeric predictor variables are *measured* rather than *manipulated* (or, in other words, having experimentally arranged conditions or groups). So regression models are, almost by definition, nonexperimental models. Even though we designate X and Y variables in a regression model, a regression coefficient is still a correlational statistic and, as discussed in Chap. 5, does not establish a causal relationship between X and Y. A regression coefficient or model statistic can only speak to the causal effect of the predictor on the outcome if (a) the predictor precedes the outcome and (b) all variables that could cause both X and Y are controlled. Since these conditions are rarely met in simple bivariate research, we should let our interpretations reflect the design in which the data were collected and use terms like "association" or "relationship" between X and Y rather than the "effect" of X "on" Y.

Table 6.4 Common conceptual elements between least squares and logistic regression

	Least squares regression	Logistic regression
Estimator	*Least squares* Regression parameters (intercept and slope) are determined by minimizing the sum of squared residuals (residual = difference between predicted Y and actual Y) from the regression line, hence *least* squares.	*Maximum likelihood* Regression parameters (intercept and slope) are selected so that the predicted y value (a predicted likelihood of the outcome) is *closest to* its actual value, hence *maximum* likelihood.
Regression coefficient	*Beta (β)* Average change in Y (in Y units) for a 1-unit increase in X 0 indicates no change in, $\beta > 0$ indicates y increase, and $\beta < 0$ indicates Y decrease, with 1-unit X increase.	*Odds ratio (OR)* Change in the odds of being in the outcome (compared to the non-outcome) category for a 1-unit increase in X OR = 1.0 indicates no change, ORs > 1.0 indicate increased odds, and ORs < 1.0 indicate decreased odds, with a 1-unit X increase.
Model fit I: Prediction error in y from x	*Residual standard error (RSE)* RSE = the standard deviation of residual scores. RSE indicates the average distance of points from regression line, or the average error of prediction. Smaller RSEs indicate better fit, and less error of prediction.	*Deviance ($-2LL$)* Log-likelihood (LL) is the sum of the "residuals"—the difference between the logged likelihoods of predicted Y and actual Y—across all X values. Smaller LLs indicate better fit, and less error of prediction Deviance is a simple arithmetic conversion of LL[a], but smaller deviance still indicates better model fit
Model fit II: Explaining y variance with x	*R^2* $R^2 = 1 - (yvar_{total} - yvar_{expl})$ R^2 indicates the proportion of total Y variance that is "explained" by X. better fitting models explain more of the outcome's variability.	*R_L^2* $$R_{\text{logistic}}^2 = -2LL_{\text{null}} - \left(\frac{-2LL_{\text{new}}}{-2LL_{\text{null}}} \right)$$ This indicates how much the intercept-only or "null" model fit is improved by X. better fitting models reduce deviance and thus explain more of the outcome's variability.

[a]-2LL allows the deviance from logistic regression models to be compared with a chi-square test, which goes beyond the scope of this course

6.8   Data Analytic Example 2

Our second example will use data we analyzed in Chap. 4 from the German Breast Cancer Group (gbsg) featuring data from a clinical trial of 686 patients with breast cancer. In this example our research question is "what is the relationship between number of positive cancer nodes on survival (measured as cancer recurrence or death before the end of the trial: yes/no)"? Because the nodes variable (*X*) numeric and *Y* is categorical, this question takes the form of a logistic statistical model.

The chunk of output below shows the following data analytic operations, in order:

- Fit the logistic regression model and generate regression coefficient, deviance statistics, R^2_{logistic} , and the point-biserial correlation coefficient.
- Label the 0,1 outcome variable for plotting purposes and generate the proportions table needed to plot the *X-Y* relationship.
- Retrieve the Pearson residuals from the model and display them with a boxplot.
- Remove the two outlier residuals from the analysis and regenerate the regression coefficient and fit statistics.

```
reg=glm(status ~ nodes,data=gbsg,family=binomial)
exp(reg$coefficients)

## (Intercept)        nodes
##   0.4573587    1.1126478

reg$null.deviance

## [1] 939.6781

reg$deviance

## [1] 895.9161

Rsq<-((reg$null.deviance)-(reg$deviance))/reg$null.deviance
Rsq

## [1] 0.04657132

cor.test(gbsg$nodes,gbsg$status)$estimate

##       cor
## 0.2422873

#we label the levels of the 0,1 variable for plotting purposes
gbsg$illness.status=factor(gbsg$status,level=c(1,0),labels=c("recurrence/death","no
recurrence"))
library(Hmisc)
gbsg$nodes3=cut2(gbsg$nodes,g=3)
tab=table(gbsg$illness.status,as.factor(gbsg$nodes3))
addmargins(tab)

##
##                   [1, 3) [3, 6) [6,51] Sum
##   recurrence/death     93     77    129 299
##   no recurrence       204    100     83 387
##   Sum                 297    177    212 686
p=prop.table(tab,margin=2)
p

##
##                     [1, 3)     [3, 6)     [6,51]
##   recurrence/death 0.3131313 0.4350282 0.6084906
##   no recurrence    0.6868687 0.5649718 0.3915094

barplot(p,col=c("blue","red"),
        ylim=c(0,1),beside=T,
        legend=rownames(tab),args.legend=list(x="topright"),
        main="Proportion of patients who experienced cancer
        recurrence or death by number of cancer nodes")
```

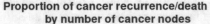

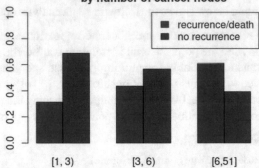

Proportion of cancer recurrence/death
by number of cancer nodes

```
#boxplot rule
res<-residuals(reg,type="pearson")
library(car)
Boxplot(res, id.method="y")

## [1] 374 541
```

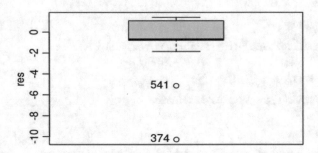

```
#refit model

reg2=glm(status ~ nodes,data=gbsg,family=binomial,subset=-c(374,541))
exp(reg2$coefficients)

## (Intercept)        nodes
##   0.4074209    1.1440292

reg2$null.deviance

## [1] 937.3838

reg2$deviance

## [1] 877.8096

Rsq<-((reg2$null.deviance)-(reg2$deviance))/reg2$null.deviance
Rsq

## [1] 0.06355373
```

6.9   Writing Up a Descriptive Analysis

A written summary of the "Data Analytic Example 2" is below.

Results

The relationship between number of positive cancer nodes and cancer recurrence (yes/no) in a sample of breast cancer patients (N = 686) was examined with a logistic regression analysis. The regression coefficient (OR = 1.11) showed that each additional node was associated with 11% higher odds of cancer recurrence compared with no recurrence. The OR and the correlation coefficient (r_{PB} = 0.24) indicate a small positive relationship between nodes and recurrence. Patients' number of positive nodes explained about 5% of the variability in recurrence. The plot below shows that as the number of positive cancer nodes increases, the probability of cancer recurrence increases. An analysis of the Pearson residuals revealed the presence of two influential cases. When the model was refit without those cases, the relationship and fit were not substantially different (OR = 1.14, R^2 = 0.06) (Box Fig. 6.1).

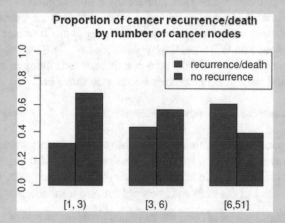

Box Fig. 6.1 Proportions of breast cancer patients who experience cancer recurrence or death as a function of the number of cancer nodes

6.10   Problems

The problems below all use datasets in the MASS package. Use library(MASS)
to make the datasets available and? datasetname *to see the documentation for
each, which will provide variable names and measurement details.*

1. In the birthwt dataset, analyze the relationship between number of previous
   premature labors (ptl) and whether the baby is low birthweight (coded 0 = not
   low, 1 = low) with a logistic regression. Generate and interpret the OR and R
   squared for the relationship. What are the predicted probabilities from your
   model of a woman having a low birthweight baby if she has had 0 and 2 previous
   premature labors?
2. In the birthwt dataset, analyze the relationship between number of physician
   visits during the first trimester (ftv) and whether the baby is low birthweight.
   Generate and interpret the OR and R squared for the relationship. What are the
   predicted probabilities from your model of a woman having a low birthweight
   baby if she has had 0, 2, and 4 physician visits in the first trimester?
3. In the Melanoma dataset, analyze the relationship between tumor thickness and
   whether the participant died from melanoma or not during the study.

   A bit of data wrangling is needed to do this problem. Look at the documentation for the Melanoma dataset and notice how status is coded. We want to create
   a new status variable (coded 0 = no and 1 = yes) because the event of interest is
   death, and we're not interested in the third category at all. To create that variable,
   run the line below. You will need the car package installed, but the car::prefix
   below will library it for you. Make sure you understand what you're doing to the
   old variable, and what you're creating, before you move on.

   ```
   Melanoma$status2=car::recode(Melanoma$status, "2=0; 1=1; else=NA")
   ```

   Run a frequency table for this new variable to see how cases are distributed in
   the two categories. Estimate a logistic regression model and interpret the OR and
   R squared for the relationship between tumor thickness and death. Generate predicted probabilities of death from the model for tumor thickness = 5 and 10 mm.
4. Using your status2 variable from the previous problem, is age of patient related
   to probability of death? Generate the OR from the logistic regression and interpret. Explain how OR = 1 the same as p = 0.5.
5. In the survey dataset, analyze the relationship between age of participant and
   whether their writing hand is left or right. To create a 0,1 outcome variable from
   the factor W.Hnd, run the following line of code. Note that this make left-
   handedness the "event" in your model.

   ```
   survey$Wrhand=car::recode(survey$W.Hnd, "'Left'=1; 'Right'=0")
   ```

Do a logistic regression and interpret the OR and R square for the relationship between age and handedness. Generate a frequency table of your outcome variable to determine if you have extremely unbalanced data? What is your conclusion?

If you had made right-handedness the event (coded 1 in your outcome variable), how would that have changed the OR? Check it by recoding the wrhand variable so that right-handedness is the event. Does the new OR describe the same relationship?

Chapter 7
Statistical Inference I: Randomization Methods for Hypothesis Testing

Covered in This Chapter
- Sample vs population and statistic vs parameter
- Hypothesis testing with a randomization test
- Monte Carlo and permutation tests
- p-values and statistical conclusion validity

7.1 Introduction to Statistical Inference

In the previous four chapters, we focused on using statistics to summarize and explore X-Y relationships in various statistical models. In that data analytic mode, we used effect size statistics to describe the direction and strength of a relationship. While descriptive analysis is important to understand relationships between variables, scientists and analysts are often interested in two further issues: Is an observed X-Y relationship "statistically significant"? And, can we use sample findings to estimate unknown quantities in the population? Both of these data analytic questions require that we shift our focus to using statistics to make *inferences* from sample data. These different roles of statistics—to describe and to infer—are bound up with the relationship between sample and population. Let's work through the dynamics of that relationship first (see Fig. 7.1).

A *population* is the complete group of participants or cases that, through your research, you would like to describe or make statements about. Often researchers refer to the *population of interest*, as it defined by characteristics or limiters that are relevant to the research question. For example, if we were analyzing 2011–2012 NHANES sample data, our population of interest would be "all Americans" because the survey is designed to describe health and nutrition behaviors in a representative sample of Americans. If our research question, however, addressed a relationship in a particular subgroup, such as people age 65 or older living in poverty, then our

© The Author(s), under exclusive license to Springer Nature
Switzerland AG 2023
B. Blaine, *Introductory Applied Statistics*,
https://doi.org/10.1007/978-3-031-27741-2_7

Fig. 7.1 The population—
sample dynamic in
statistical inference

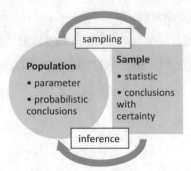

population of interest would be older Americans living in poverty. A *sample* is a subset of the population of interest from which you collect data for research. Samples are often only a very small piece of the population, but as we covered in Chap. 2, sample size is less important to statistical inference than whether the sample is a random subset of the population or not.

Research is generally done in, or with, sample data, and any research question that can be addressed by sample data can be answered *with certainty*—for the sample. For example, in the 2011–2012 NHANES sample data (N = 5000), we can *know* the proportion of female participants in the sample, the mean BMI score for the sample, or any other statistical summary of a variable in the study. We can't know (not with certainty) the proportion of female participants or the mean BMI score in the population from which the NHANES sample came (i.e., all Americans). Numbers that describe populations are called *parameters*. Parameters are generally unknown because research is done with sample rather than population data. Furthermore, random samples differ, and the particular sample of 5000 Americans from 2011 to 2012 is just one of a very large number of possible samples that could have been drawn. From the population of all Americans, countless random samples of 5000 can be drawn, and each will be a unique subset of the population. Each of those samples will produce a different "answer" to the research question, and we have no idea how close *our* sample's "answer" is to the correct answer (the correct answer being the value of the parameter) is. As a result, conclusions made about the population from sample data are uncertain. Inferential statistics help us quantify that uncertainty with probabilities, allowing us to make probabilistic conclusions about population values or statements (e.g., the mean BMI score for all Americans). These probabilistic conclusions are called *estimates*.

In this dynamic relationship between population and sample, the ultimate goal of research is to *know* population parameters and characteristics. Given that we only have samples and sample data to work with, the process of developing accurate conclusions about population parameters is a long-term project. Nothing is ever "proven" or known with certainty by the evidence in a single study. Rather, statistical inference is a disciplined process of accumulating and evaluating statistical evidence from sample data to make increasingly confident assertions about the state of affairs in the population.

In Chap. 2 we learned that the conclusions we make about generalizability (i.e., do sample-based findings generalize to the population?) and internal validity (i.e., does *X* cause *Y* or are they merely correlated?) are *inferences* about things that exist in the population. Although those inferences weren't explicitly probabilistic or require statistical methods, it is worth remembering that developing accurate knowledge (i.e., highly probable inferences) about population relationships requires evidence from more than one sample or study. In this chapter we learn about a third dimension of inference called statistical conclusion validity. *Statistical conclusion validity* refers to confidence that an effect or relationship observed in sample data is not simply a random event but reflects a true relationship in the population. As we observed earlier, millions of random samples can be taken from a population, and their estimates of some *X-Y* relationship will differ. We therefore need a method for evaluating the relationship in *our* sample data against the range of outcomes than occur randomly. If our sample finding is an unlikely event in that range of random outcomes, we want to attach a probability to that conclusion. We now turn to the focus of this chapter—how to use randomization methods to assess statistical conclusion validity. In the next chapter (Chap. 8), we will continue our use of statistical inference methods to address parameter estimation.

7.2 Randomization Test

How do we use statistics to develop statistical conclusion validity? In this section we learn the basic elements of a *randomization test*, through a simple example, and then extend those elements to understand Monte Carlo and permutation tests, which are variations on the randomization test. All three use resampling methods to do statistical inference. *Resampling methods* repeatedly sample from study data to develop distributions of outcomes that enable statistical inference. Let's see how resampling is used in the randomization test.

Imagine we have a brood of six newborn chicks and we want to test the effect of a special diet on weight. Because a randomization test depends on random assignment, let's assume the chicks are randomly assigned to diet condition (normal or special) in the study and after 6 weeks their weight is measured in grams (g). The study results show that the chicks on the normal diet weighed 136 g, 141 g, and 179 g (mean weight = 152 g) and the chicks on the special diet weighed 160 g, 181 g, and 229 g (mean weight = 190 g). The basic question of statistical conclusion validity in this study is: Is this difference between mean weights of 152 g and 190 g random, or might the difference reflect an effect of diet on weight?

Is there evidence that the diet had an effect on chick weight? First let's acknowledge that if you had a large population of 6-week-old chicks, the mean weights of random samples of three chicks would differ: some groups of chicks would be lighter or heavier than others—just by chance—because weight varies. So mere sampling variation could explain any outcome, including our study's outcome.

What we need to know is *how likely* it is for two randomly arranged samples of three chicks to have mean weights as different as (or more different than) 152 g and 190 g.

The randomization test was developed by Sir Ronald Fisher in the 1920s to evaluate the null hypothesis associated with some treatment or intervention. A null hypothesis (H_o) is a statement that the treatment (X) has no effect on the outcome (Y). In our example, H_o would be "Diet has no effect on chick weight." The alternative hypothesis (H_A) would include all other outcomes. In our example, H_A would be "Diet has an effect on chick weight." The logic of a randomization test starts with assuming that the null hypothesis is true and rests on the *principle of exchangeability*. If H_o is true (and there is really no effect of special diet compared to normal diet on weight), then the chicks' weights are exchangeable—that is, any weight value observed in the normal diet group could just as easily have shown up in the special diet group and vice versa. Under the null hypothesis then, we should be able to simply randomize our 6 chick weights into 2 groups of 3 scores each because all 6 scores are exchangeable. A randomization study produces all possible random arrangements of N scores into groups of k scores (in our study, $N = 6$ and $k = 3$). For you math nerds, the number of permutations of N scores into samples of size k and $N-k$ can be calculated as follows:

$$\binom{N}{k} = \frac{N!}{k!(N-k)!}$$

(7.1)

For each permutation (see footnote "a" in Table 7.1) or random arrangement of scores, the statistic of interest (see footnote "b" in Table 7.1) is calculated. That procedure generates a *reference distribution* of the statistic (in this case we're interested in the mean difference) under H_o. Table 7.1 below shows the 20 permutations of 6 scores into 2 groups of 3 from our example, as well as the group means and mean difference for each. Notice that our study data (highlighted below) is one of those random arrangements.

Next we convert our observed statistic, which is often referred to as a *test statistic*, to a probability. We use the reference distribution from the randomization procedure and find the probability of an outcome that is at least as large as our observed outcome (test statistic $= -38$ g). In Table 7.1 we see 4 scores ≥ 38 g (in the positive or negative direction), so the probability of observing a mean difference ≥ 38 g is 4/20, or 0.20. Why do we disregard the sign of the mean difference when computing the probability? Because a randomization test is a test of H_o, and the alternative to "no effect" is "any effect." This probability (written formally as p(observed | H_o) $= 4/20 = 0.2$) is called a *p-value*. The p-value is a *conditional* probability, with the vertical line in the statement above indicating "conditioned upon" or "given." So, conditioned upon H_o being true, the probability of observing a mean difference as large or larger (in either direction) than the difference we observed in our study is 0.2.

Table 7.1 Permutations of sample data where $N = 6$ and $k = 3$

Permutation	Group 1	M_1	Group 2	M_2	$M_1 - M_2$
1	136 141 160	145.7	179 181 229	196.3	-50.7
2	136 141 179	152	160 181 229	190	-38
3	136 141 181	152.7	160 179 229	190	-36.7
4	136 141 229	168.7	160 179 181	173.3	-4.7
5	136 160 179	158.3	141 181 229	183.7	-25.3
6	136 160 181	159	141 179 229	183	-24
7	136 160 229	175	141 179 181	167	8
8	136 179 181	165.3	141 160 229	176.7	-11.3
9	136 179 229	181.3	141 160 181	160.7	20.7
10	136 181 229	182	141 160 179	160	22
11	141 160 179	160	136 181 229	182	-22
12	141 160 181	160.7	136 179 229	181.3	-20.7
13	141 160 229	176.7	136 179 181	165.3	11.3
14	141 179 181	167	136 160 229	175	-8
15	141 179 229	183	136 160 181	159	24
16	141 181 229	183.7	136 160 179	158.3	25.3
17	160 179 181	173.3	136 141 229	168.7	4.7
18	160 179 229	189.3	136 141 181	152.7	36.7
19	160 181 229	190	136 141 179	152	38
20	179 181 229	196.3	136 141 160	145.7	50.7

[a]In math, permutations imply different orderings of objects, but in statistics we use the term synonymous with random arrangement. We don't consider the same scores but in a different order to be a unique random arrangement, mainly because they produce the same statistic

[b]The *statistic of interest* will depend on (a) the statistical model you're working with and (b) the particular statistical "lens" that you're using to study the relationship. In this example, we're working with ANOVA model data and using the mean difference to study the effect of diet on weight in chicks. A randomization test can be done with any statistic and with data from any of the four models covered in Chaps. 3, 4, 5, and 6

7.3 Interpreting the Results of a Randomization Test

Interpreting the results of a randomization test hinges on three principles: proper understanding of a p-value, whether the assumption of exchangeability is met in the original study data, and what statistical alternatives are included in the alternative hypothesis. Let's take each of these principles, think them through, and see how they affect conclusions from a randomization test.

7.3.1 p-Value

To understand the meaning of, and to properly interpret, a p-value, we use the guidelines set out by Sir Ronald Fisher.

- *A p-value is a continuous measure of evidence against H_o.* The lower the p-value, the stronger the evidence against H_o.
- *A p-value is just one piece of evidence.* A p-value must be combined with other pieces of evidence, ideally across multiple studies, to help draw conclusions from data. A p-value from a single study doesn't "prove" anything: a low p-value is evidence against the validity of H_o but does not establish H_o as false.
- *A p-value is not evidence of H_A.* A low p-value does not speak to the validity of H_A or establish it as true. Remember, if H_o is false, there are many alternative hypotheses to be considered as potentially responsible for the observed outcome. And, again, evidence from multiple studies is needed to validate a particular H_A.
- *A p-value of 0.05 is a reasonable cutoff.* According to Fisher, $p < 0.05$ is a reasonable cutoff for evidence that calls the validity of H_o into question and allows us to further consider the alternative hypotheses. To interpret a p-value, we also need to acknowledge that it is a continuous measure of evidence, so some interpretive rules might help.

p value	Evidence against H_o
>0.05	None
0.01–0.05	Weak
0.001–0.01	Moderate
<0.001	Strong

We referred earlier to scientists' interest in learning if a particular *X-Y* relationship is "significant" and may have heard the terms "significant," "statistically significant," and "nonsignificant." These terms are ubiquitous in scientific articles, as well as media coverage of science, and are used to characterize study findings where the p-value was below some threshold value set to declare a finding significant. We can declare a finding with a small p-value "significant" or "statistically significant" but we should do so mindful that a p-value is a continuous measure of evidence. When a p-value is used to confer a binary status ("significant" or "nonsignificant"), we abandon the true continuous nature of the underlying evidence. Declarations of "significance" based on a p-value also convey much more certainty about the outcome than is warranted, or that Fisher ever intended in setting up the principles of statistical inference. Second, the term "significant" in common usage often implies "meaningful" or "important." A p-value however is merely a probability of some observed outcome (or greater) occurring under the null hypothesis; it says nothing about the importance of the outcome.

Our randomization test of the effect of special diet compared to normal diet on weight in six chicks produced a p-value of 0.2. This indicates that there was no evidence against the null hypothesis, *in that particular study*. Remember, a p-value from single study does not confirm H_o (i.e., concluding that there is absolutely no effect of special diet on chick weight in the population), so another study with different chicks, different bags of feed, or different environmental conditions might find something different.

7.3.2   Exchangeability

A randomization test assumes exchangeability in the original scores, meaning that all random arrangements of scores are equally likely. This is an important assumption because the randomization test gives equal weight to each permutation of scores in the calculation of the p-value. If all permutations of scores are *not* equally likely, then the p-value is not accurate and mischaracterizes the amount of evidence against H_o. When the data come from an experiment, the assumption of exchangeability is justified. Indeed, a randomization test mirrors the random assignment of participants (i.e., chicks) to condition (i.e., special vs normal diet) in a study. In that case, as we learned in Chap. 2, causal conclusions can be made about randomization test results. Although our chick study did not result in much evidence against H_o, if it had, we could have interpreted that difference in weights as caused by the diet variable. In the next section, we look at how a randomization study changes when the assumption of exchangeability is not justified.

7.3.3   Statistical Alternatives Under H_A

As stated earlier, the alternative hypothesis (H_A) in a randomization test is not a specific hypothesis; it is an umbrella hypothesis that includes all "non-null" outcomes. This is why our H_A was stated "Diet has an effect on chick weight." Implicit in that statement of H_A however is the following acknowledgment: variables that confound diet are also part of that umbrella of alternatives. So if our study had generated a small p-value, we still couldn't be certain about what *did* produce the outcome (see bullet #3 in 7.3.1). It is possible that diet caused the outcome, but it is also possible that some variable or combination of variables that covaried with diet is responsible. This points up the importance of controlled study designs (like experiments) for causal conclusion validity; controlled studies reduce the number of alternative explanations for the observed relationship.

It is important to recognize that a treatment or intervention can be reflected in location statistics (e.g., the mean difference) or variability statistics (e.g., variance ratio). A randomization test can be conducted using any outcome statistic, but interpretations of the test are limited to that statistic. If, for example, we expect that feeding chicks a special diet will affect all chicks about the same, then a location statistic like the mean difference would capture that effect. If we suspected however that the special diet in particular would make some chicks heavier than others, then a variability statistic might better capture that effect. Our interpretations of randomization test results, then, should acknowledge the particular statistical lens used to view the outcome. With that said, we learn more about an intervention's effect on an outcome if we view it through different statistical lenses. We will demonstrate this below, as well as in Chap. 9, when we use randomization tests in data analysis.

7.4 Subtypes of the Randomization Test

The randomization test has two variants: the *Monte Carlo test* and the *permutation test*. Although they are all often referred to as "randomization tests," it is important to know how they differ because a Monte Carlo test involves different computations, and a permutation test different interpretations, than the randomization test covered above. We demonstrate these two types of randomization studies in in Chap. 9, but for now let's distinguish them conceptually.

7.4.1 Monte Carlo Test

A true randomization test uses *all possible random arrangements* of scores to construct the reference distribution of the statistic, which in turn produces an *exact* p-value. In most studies, "all possible random arrangements" is a very large number and often too large to be computationally feasible. For example, in a study that randomly assigned 50 participants to one of two groups (25 per group)—which is really not a large study as studies go—there are 1.26×10^{14}, or 126 trillion, random arrangements of scores. It would take most desktop computers several hours to work through the calculations of such a large randomization test. Monte Carlo methods operate on the same principle as random sampling: a random sample of sufficient size generates an accurate picture of the population, allowing sample-based findings to be confidently generalized to the population. In this case we're interested in accurately capturing the true reference distribution, so a Monte Carlo test randomly samples from the population of all random arrangements to generate an approximation of the true reference distribution. The p-value produced by a Monte Carlo distribution of a statistic is not an exact p-value, because only a subset of all possible random arrangements is included in the distribution. Monte Carlo samples of around 5000 generate estimated reference distributions that are very accurate and provide estimated p-values that are very close to exact p-values. Because very few studies are as small as our chick weight study (N = 6), most randomization studies actually use Monte Carlo methods, generating estimated rather than exact reference distributions and p-values.

7.4.2 Permutation Test

A true randomization test assumes exchangeability in the original study data. This is assured by random assignment of participants to conditions (i.e., experimental research), in which scores from one condition could have just as easily been in the other condition. In studies where the predictor variable is selected rather than experimentally arranged (i.e., non-experimental research), scores from one condition are *not* exchangeable with scores from the other condition. In Chap. 3 we compared female and male NHANES study participant scores on systolic blood pressure. Female and male American adults are already different when they enter the study, so their scores are not exchangeable. The lack of random assignment in the original study does not preclude the use of randomization methods for hypothesis testing. A permutation test, then, is simply a randomization study in data from a study in which exchangeability did not hold. A study that doesn't randomize participants to groups cannot establish exchangeability. Nevertheless, a permutation study still produces a reference distribution (of, say, the mean difference between females and males on blood pressure) which in turn generates a p-value (exact if true randomization, estimated if Monte Carlo) that speaks to the null hypothesis. However, causal conclusions about the "effect" of gender on systolic blood pressure are precluded.

7.5 Doing Randomization Tests in R

In this section we learn how to set up and run a randomization test in R. The focus will be understanding the logic of the coding needed to carry out the resampling procedure and generating and using the reference distribution to do null hypothesis testing. The goal is that once you understand the basic architecture of the randomization test as it is expressed in R code, you can adapt it to address any hypothesis testing problem with any statistic.

7.5.1 Example 1: Randomization Test Using the Mean Difference

We will use the Melanoma data in the MASS package for our example, which is data on 205 patients with malignant melanoma. Let's concern ourselves with the question of the relationship between *ulcer* status (X, coded 1 = presence, 0 = absence) and survival *time* (Y, measured in days). Because we have a categorical predictor and a numeric outcome, and thus an ANOVA statistical model, refer to Chap. 3 for reminders on how to do descriptive data analysis. To prepare for the randomization study, we need to do a couple key bits of descriptive analysis, generating mean survival times by ulcer group, group sample sizes, and boxplots of the outcome by group (see output below). Each of these will become apparent as we work through the example.

```
library(MASS)
library(lattice)

tapply(Melanoma$time,Melanoma$ulcer,mean)

##        0        1
## 2414.965 1817.811

table(Melanoma$ulcer)

##
##   0   1
## 115  90

bwplot(ulcer~time,data=Melanoma
```

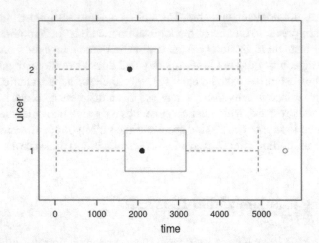

The randomization test (actually, in this case it's a permutation test with Monte Carlo methods) tests the null hypothesis, so let's set up our hypotheses for this test.

H_o: There is no relationship between ulcer status and melanoma survival time.
H_A: There is a relationship between ulcer status and melanoma survival time.

Notice that I used the phrase "relationship between" rather than "effect on" in the hypothesis statements. This recognizes that ulcer groups were not experimentally arranged in the original study. Rather, melanoma patients developed an ulcer or not (or they already had an ulcer prior to the study or not), and those with an ulcer were compared to those with no ulcer. We can imagine those groups of patients differed in many other ways than just the presence or absence of an ulcer. Lacking experimental manipulation of the predictor variable in the original study (which in this case would involve randomly assigning people to be given an ulcer!), we cannot infer any causal effect of ulcer on survival time. These observations about the study design also confirm the lack of exchangeability in the data, requiring a permutation study. We will be using Monte Carlo methods because the number of unique random arrangements (or permutations) of 205 scores into groups of 115 and 90 is on the order of 10^{58} and not computationally feasible. Let's walk through the randomization test in steps:

Step 1. Set Up Test Statistic The first step in the randomization (technically, here, a Monte Carlo permutation) test is to decide on the outcome statistic we will use to test H_o. The decision depends somewhat on how you think the predictor will affect the outcome and what's the best statistical lens to "see" that effect. Let's use the mean difference as our test statistic. Below we find the mean difference in survival time between ulcer conditions and save it in an object (*obs*) for later use. Based on the group means above, we also arranged the calculation so that the mean difference is a positive number; the reason for that will be clear in Step 4.

```
obs=(mean(Melanoma$time[Melanoma$ulcer == 0]))-(mean(Melanoma$time[Melanoma$ulcer
== 1]))
obs
```

```
## [1] 597.1541
```

The mean difference in survival time between patient with and without an ulcer is about 597 days. We want to know how probable it is to get a difference that large or larger if H$_o$ is true, or in other words if having an ulcer is actually unrelated to survival time.

Step 2. Set Up Resampling Loop   The next step in the randomization study is to set up the loop by which we will sample from the sample data, randomize the sample values into groups, calculate the statistic of interest, store that value, and repeat the process. We also want to make sure that our loop conducts this process under the condition that H$_o$ is true. Below is a loop set up to conduct that resampling task. Since this introduces some new R functions and concepts, let's walk through the code line by line.

```
set.seed(1008)
N=5000
meandiff=numeric(N)

for (i in 1:N) {
  data <- sample(Melanoma$time,205,replace=FALSE)
  meandiff[i] <- mean(data[1:115])-mean(data[116:205])
}
```

- The set.seed() function is only there to ensure that when you run the code you get the same results as everyone else. Anytime we use the `sample()` function, it will produce independent and unique sets of objects for each person. So for this example only, I wanted everyone to end up with the same results, hence the set.seed.
- N defines the number of times we run the set of operations in the loop. For a Monte Carlo study, the number of samples needed to create an accurate reference distribution is recommended to be 5000.* Please note that N is also commonly used in statistics as symbol for sample size.
- The `numeric()` function creates a numeric vector, called *meandiff* in this example, which will eventually contain N values. Think of it as a container in which you're storing each statistic generated by one pass through the resampling procedure.
- The `for()` function is the engine of a for-loop, which is a repeating series of operations that result in a piece of output. Each pass through the loop is an i (or iteration), and we pass through the loop N times. For each i, we:
  - Take a random sample of our time data. Because we are sampling *without* replacement (replace = FALSE), the `sample()` function will select a ran-

dom time value, store it in the object, then select another random value from the remaining 204 cases, etc. until all 205 cases have been selected. In other words, sampling without replacement essentially creates a random *ordering* of the 205 values in the time variable. We store that random arrangement of values in an object called *data*.

- Divide the 205 values into two groups of the size of the original ulcer groups. It is important to maintain the original group sample sizes in a randomization study. This process is no different than shuffling a 205-card deck and dealing the cards into piles of 115 and 90. The randomizing of time values into groups is what we expect to happen under H_o: group 1 and group 2 will differ, but only randomly. Calculate the mean difference from our randomized groups and deposit that value in the numeric *meandiff* container created above.

This process repeats N (or 5000) times, each time with a new random sample of 205 survival time values.

Step 3. Generate Reference Distribution The reference distribution is the distribution of mean difference statistics for a random sample of 5000 permutations of 206 scores into groups of 115 and 90. We don't need to look at a histogram of the reference distribution to use it to find our estimated p-value, but the histogram helps us to see and appreciate how the outcome statistic—which in this case is the mean difference—behaves under H_o. We can see that most sample mean differences cluster around zero, which is logical if we remind ourselves that under H_o ulcer status is unrelated to survival time. Larger mean differences in both directions are progressively less probable occurrences under H_o.

```
hist(meandiff, main="reference distribution of survival time
mean difference under Ho")
```

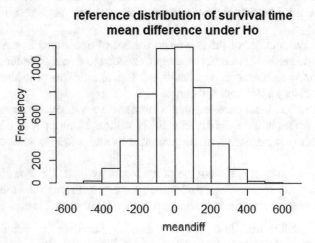

reference distribution of survival time mean difference under Ho

Step 4. Calculate p-Value The p-value for our example is the probability of observing a mean difference at least as large as our test statistic, under H_o. Since all the outcomes in the *meandiff* container occurred under H_o, we simply need to count those outcomes that exceeded 597.2 and divide by N to get the probability. Remember that we want to count values that are larger than the test statistic in both the positive and negative direction, because H_A includes the alternative hypotheses that ulcer, compared to no ulcer, will be associated with both less and more survival time. In R we can do this by taking the absolute value (with the abs() function) of the resampled mean differences and using the length() function to count how many are greater than or equal to the test statistic and then divide that total by N.

```
pvalue=length(which(abs(meandiff)>=obs))/N
pvalue

## [1] 2e-04
```

Having converted our test statistic to a probability under H_o, we find p = 0.0002. Remember, this is an estimated, not an exact, p-value; the accuracy of the estimate is tied to N. We could have set N higher, say to 10,000, and we might not have even noticed the difference in computing time to run the study. However, research shows that the reference distribution produced by an N = 10,000 randomization study wouldn't be any more accurate, and thus wouldn't deliver any more accuracy in the estimated p-value, than a N = 5000 study. This p-value constitutes strong evidence against H_o, encouraging us to explore alternative hypotheses. The data from the original study suggest that the presence of an ulcer in melanoma patients is associated with shorter (rather than longer) survival times. This is a plausible conclusion, but it is still one of many plausible alternative hypotheses, mainly because the original study was not experimental and thus could not eliminate those alternative explanations with random assignment of patients to ulcer condition.

7.5.2   Example 2: Randomization Test Using the Median Difference

Let's look at one more example of a randomization test by using the median difference as our outcome statistic. Referring back to the boxplots and descriptive statistics generated earlier, we see two things: one, the presence of an outlier in the high end of the no-ulcer group, which has an effect on the mean of that group, and, two, the group medians appear to be less different than the group means were. So, as a way to enhance statistical conclusion validity with regard to the conclusion that *typical* (not just mean) survival times are higher in patients with no ulcer compared to patients with an ulcer, let's test that hypothesis with a different location statistic—the median difference. H_o and H_A remain the same in this test.

We combined all the steps of the randomization study for this example, but the changes to the code are noted below with #annotations.

```
#new test statistic
obs2=(median(Melanoma$time[Melanoma$ulcer == 0]))-
  (median(Melanoma$time[Melanoma$ulcer == 1]))
obs2
```

```
## [1] 303.5
```

```
N=5000
meddiff=numeric(N)        #newly named numeric vector object
set.seed(1008)
for (i in 1:N) {
  data <- sample(Melanoma$time,205, replace=FALSE)
  meddiff[i] <- median(data[1:115])-median(data[116:205]) #calculate median
}
hist(meddiff, main="reference distribution of survival time
      median difference under Ho")
```

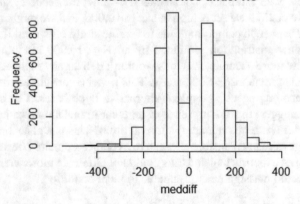

```
pvalue=length(which(abs(meddiff)>=obs2))/N
pvalue
```

```
## [1] 0.012
```

Notice that the median difference (303.5 days) is roughly half of the size of the mean difference. Nevertheless, that difference is still an improbable event under H_o ($p = 0.012$), constituting weak to moderate evidence against H_o. Together, these two randomization studies—each with a different location statistic—provide consistent evidence that the observed differences in *typical* survival times of ulcer and no-ulcer patients are not random. They also illustrate that statistical conclusion validity is best established with an accumulation of evidence, both within a study and accumulated across studies.

7.6   Writing Up the Results of a Randomization Test

We will do much more with randomization tests in data analysis in Chap. 9, where we use randomization methods to testing hypotheses in data from each of the four statistical models covered in previous chapters. We will also continue to examine how the principle of exchangeability affects interpretations of randomization studies. To end this chapter, here is a written summary of the analysis we did on the ulcer and survival time data. Just a reminder: we don't know how these patients were sampled, so we can't assume random sampling, and thus these findings lack external validity. We established earlier that the lack of random assignment, and exchangeability, in the original study weakens the causal inference we might have made about the *effect* of having an ulcer on survival time.

Results

The relationship between the presence and absence of an ulcer on survival time (in days) was examined in a sample of patients with malignant melanoma (N = 205) with a randomization test of the median difference. The median survival times for patients with, and without, an ulcer were 1799.5 days and 2103.0 days, respectively (see Box Fig. 7.1). The randomization test created a reference distribution of 5000 median differences under the null hypothesis of no relationship between ulcer status and survival time. The observed median difference (303.5 days) had an associated p-value of 0.012. This constituted weak to moderate evidence against H_o and suggests that the presence of an ulcer is associated with shorter survival times in melanoma patients.

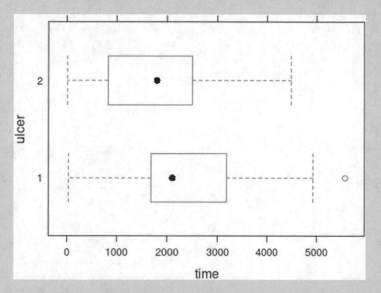

Box Fig. 7.1   Survival time (days) by ulcer group (0 = absence, 1 = presence)

7.7   Problems

1. What dimension of statistical inference is related to:

    (a) Sampling in the original study?
    (b) Whether the original study is an experiment?
    (c) Distinguishing between random and non-random study outcomes?
    (d) Estimating the true magnitude of an effect or relationship?
    (e) Showing that study outcomes are reliable?

2. What is the assumption of exchangeability and how is it consequential in randomization studies?
3. Describe the difference between a randomization study and a Monte Carlo study?
4. What is a reference distribution and what role does it serve in hypothesis testing with a randomization test?
5. Can exchangeability be assumed in data from non-experimental studies? Explain why not.
6. What is the difference between the null and alternative hypotheses in a randomization test?

Chapter 8
Statistical Inference II: Bootstrapping Methods for Parameter Estimation

Covered in This Chapter

- The logic of bootstrapping
- Parameter estimation and confidence intervals
- Bootstrapped confidence intervals
- Bootstrapping for estimation in R

8.1 Introduction to Parameter Estimation

Statistical inference involves making probabilistic conclusions, based on sample data, about population parameters and relationships between variables. In Chap. 7, using the methods of randomization and randomization tests, we assessed the statistical conclusion validity associated with an *X-Y* relationship. Statistical conclusion validity (i.e., having strong evidence against H_o) does not, however, speak to the *magnitude* of that effect or relationship. This is an important issue because statistical conclusion validity can occur even when the effect size is trivially small, if certain conditions are present (e.g., large N, small sample variances, or a combination of both). So "statistical significance" does not imply that the observed effect or relationship is large or meaningful in practical terms. We also should be aware that treatment effects or relationship sizes, as captured with effect size statistics, will vary across hypothetical identical studies, because samples vary. *Effect size* estimation is the statistical inference task that involves making probabilistic statements about the direction and magnitude of an effect or relationship based on sample data.

The *true* size and direction of an effect or relationship—meaning the size and direction of the effect in the population, could it be known—is a *parameter*. Establishing effect size validity then is a process of estimating that parameter based on sample data. In this chapter we will learn the statistical tools and methods for parameter estimation, including a resampling method called *bootstrapping*.

© The Author(s), under exclusive license to Springer Nature Switzerland AG 2023
B. Blaine, *Introductory Applied Statistics*,
https://doi.org/10.1007/978-3-031-27741-2_8

Bootstrapping is to parameter estimation as the randomization test is to null hypothesis testing. Bootstrapping allows us to create a reference distribution of outcomes—called a bootstrap distribution—under some alternative hypothesis. Recall that a p-value from a randomization test only tells us that a sample finding is unlikely under the null hypothesis. We need to figure out what "non-null" or alternative hypothesis does account for the sample finding, and bootstrapping will allow us to construct a probabilistic estimate of the parameter of that alternative distribution.

8.2   The Logic of Bootstrapping

The logic of bootstrapping operates on the law of large numbers, which states that, for any statistical experiment (e.g., a coin toss, a dice roll), the mean of a large number of repeated experiments is very close to the expected value of the outcome. For example, we know that the true probability (or expected value) of tossing heads on a coin is 0.5. If you toss a coin 10,000 times, the number of heads will be very close to 0.5. Now, if we make the statistical experiment a random sample and the outcome a statistic (e.g., mean, median), the law of large numbers says that if we repeatedly sample from a population, each time calculating a statistic, the mean of those sample statistics will be very close to the statistic's expected value. The problem is that we rarely have access to a population from which to sample, and that's where bootstrapping comes in.

In *bootstrapping* we treat our sample data as a surrogate, or stand-in, population. Indeed, *any* sample is a surrogate of the population from which it came, although as we learned in Chap. 2 random samples—particularly large random samples—are more representative of their parent populations than convenience samples, and thus much better surrogates. The primary assumption underlying statistical inference by bootstrapping is that our sample is reasonably representative of the parent population. Bootstrapping involves *sampling with replacement* from our surrogate population (the sample data), calculating and saving a statistic from that sample, and repeating the process a large number of times. Sampling with replacement means that a value is chosen from the data, recorded, and then "thrown back" into the sample. As a result, bootstrapped samples can select the same value multiple times, and that's OK. In fact, sampling with replacement is what enables bootstrapping to create samples that reflect the surrogate population. Bootstrapping uses the *plug-in principle*, which states that each bootstrapped sample statistic "plugs in" for the parameter—the unknown population value of interest. This resampling process generates a distribution of bootstrapped statistics, called a bootstrap distribution, which will enable us to estimate the parameter. These principles are illustrated in Fig. 8.1.

It is important to recognize that a bootstrapped distribution is always centered around the sample statistic, not the population parameter. Bootstrapping isn't a method for magically "seeing" the population parameter from our sample data, nor will it help us improve the accuracy of a parameter estimate. It bears repeating: the best parameter estimates come from large, random samples from the population. The accuracy of any estimate depends on the how closely the sample reflects the population—if your sample is biased, its estimates will be biased. What

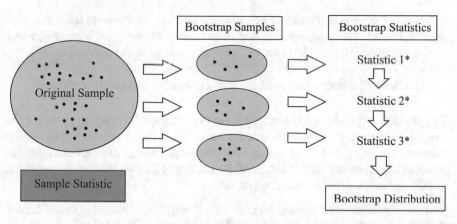

Fig. 8.1 Bootstrapping. (Source: https://www.statisticshowto.com/bootstrap-sample/)

bootstrapping *does* do is tell us how accurate, or precise, our sample-based estimate is. A bootstrap distribution approximates the true sampling distribution of a statistic; this is good because we rarely can sample repeatedly from the population and get a picture of that "true" sampling distribution. So bootstrapping is very useful for getting estimates of standard errors of statistics and creating confidence intervals. More about these concepts next.

8.3   Parameter Estimation

Let's revisit our Chap. 3 example for a minute. In that study we compared samples of boy and girl kindergarten students on a math achievement test. The point of doing the study was to be able to say something—based on that particular sample of 200 students—about the *true* difference between girls' and boys' math achievement (meaning in the population of *all* kindergarten students). Research is almost always interested in parameters, and parameters are almost always unknown, hence the need to estimate them. We are aware however that another sample of 200 kindergarten students would yield a different estimate, a third sample yet another estimate, and so on. Parameter estimation must address this problem—that sample estimates of a parameter vary—and it does so with a confidence interval.

A *confidence interval* is an interval, defined by lower and upper values, that we have some confidence includes the parameter. In interval estimation, "some confidence" is a precise value that is set by the analyst. The confidence level of a confidence interval is often set at 0.95, by convention. A 95% confidence interval, then, is an interval of values that we are 95% confident includes the parameter. We will come back to the issue of interpreting a confidence interval in a bit. For now, we need to understand how a confidence interval is constructed and where the quantities that contribute to the calculation of a confidence interval come from.

There are three contributors to a confidence interval, as shown in the conceptual formula below: a statistic, the standard error of that statistic, and a multiplier that establishes the confidence level for our interval estimate of the parameter.

$$\textbf{statistic} \pm \left(\textbf{standard error of the statistic}^* \textbf{confidence level multiplier}\right)$$

The product of the standard error and the confidence level multiplier produces the *margin of error*; a statistic is reported a lot in fields that use survey and polling research. The margin of error is then subtracted from, and added to, the statistic to generate the lower and upper values of the interval. Next we look more closely at each of the terms in the confidence interval.

Statistic Parameter estimation begins with a statistic, obviously. Our Chap. 3 example was interested in the mean difference between populations of girls and boys on a math achievement test. So we could simply use a single value—the mean difference from our sample data—to estimate the parameter; this is called *point estimation*. The value of a point estimate from a large random sample is hard to overstate. Point estimation does not account for the error inherent in any sample statistic, however. Sample statistics vary depending on the particular sample you happen to get in your study. Point estimates don't tell us how much they vary, or could be expected to vary, across hypothetical samples—you know, all the samples you *didn't* get in your study. Interval (unlike point) estimates do incorporate an estimate of the statistic's error, which is what we turn to next.

Standard Error The standard error of a statistic is the average amount of variability in the statistic when observed over many random samples of the same size from the population of interest. Because that variability is due to the random differences in samples, it is called error. But the standard error is conceptually equivalent to the standard deviation, which you recall from Chap. 2 quantifies the average deviation from the mean in a sample of scores. The standard error, then, is simply the standard deviation of a distribution of statistics. Every statistic has a standard error. The standard error of a statistic tells us how much variability, on average, we could expect if we repeatedly drew samples from the population and calculated the statistic in each. So the standard error is essential to interval estimation because it quantifies the uncertainty inherent in using a particular sample statistic to estimate the parameter, knowing that another sample would produce a different value of the statistic.

If we have the statistic, how do we find the standard error of the statistic? Well, there are two ways. First, some standard errors can be calculated from a formula. For example, the standard error of the mean ($SE_{\bar{x}}$) can be calculated from the sample standard deviation and sample size as follows:

$$SE_{\bar{x}} = \frac{s^*}{\sqrt{n}}.$$

This formula produces only an *estimated* standard error of the mean, and the accuracy of that estimate depends on assumptions (i.e., that the data are randomly sampled and normally distributed in the population) that are often not met in real data. An inaccurate standard error estimate will then pass its inaccuracy on to the confidence interval. The second method is to bootstrap the standard error of the statistic whose parameter we are interested in estimating, which we cover in the next section. We *must* use a bootstrapped standard error if we want to estimate with a statistic that doesn't have a derived standard error formula. We *can* use bootstrap standard errors for estimating with means and proportions (statistics that have standard error formulas), and we might actually want to if the assumptions underlying those formulas are not met.

Confidence Level Multiplier Let's go back to a concept you learned in high school called the *empirical rule* (see Fig. 8.2). The empirical rule applies the properties of the standard normal distribution (i.e., a normal distribution with $\mu = 0$ and $\sigma = 1$) to some random variable, which ideally is also normally distributed. The empirical rule tells us the percentage of observations in that random variable in reference to common standard deviation (or z score) cutoffs. For example, about 68% of the observations of a normally distributed random variable are expected to be between $z \pm 1$. If you turn that around to create a tool for estimating μ, you can see that the interval formed by $z \pm 1$ includes the *most likely 68% of the values* of μ. In other words, the multiplier 1*SD produces a 68% confidence interval for estimating the parameter. If we want a more confident estimate, say 95% confidence, we would need an interval that included the most likely 95% of the values of the parameter, or 2*SD. So the multiplier is just the value of the standard (in this case, z) score that sets the confidence level of the interval estimate by including that particular percentage of the most likely values of the parameter (in this case, μ).

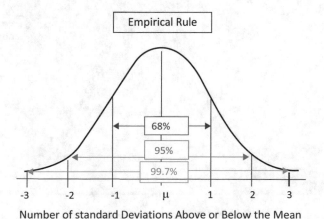

Fig. 8.2 The empirical rule. (https://www.softschools.com/math/probability_and_statistics/the_normal_distribution_empirical_rule/)

Normal-Theory Interval Estimation Let's apply these ideas to the problem of estimating a population mean (μ) from sample data using the classic normal-theory method. Below is the conceptual formula introduced earlier for building a confidence interval but represented in terms for building a 95% confidence interval to estimate the population mean.

$$\bar{X} \pm \left(SE_{\bar{X}} * t_{.95} \right), \text{where}$$

$\bar{X}$ = sample mean

$SE_{\bar{X}} = $ standard error of the mean, which is estimated by $\dfrac{s}{\sqrt{n}}$

$t_{.95}$ = multiplier associated with 95%

Notice two things: First, our estimate of μ begins with $\bar{X}$. The sample mean is a reasonable place to start to estimate the population mean. In fact, when we use the sample mean to estimate the population mean, it is called a *point estimate*. The drawback of a point estimate however is that it provides no clue as to the error inherent in using a sample statistic to estimate the corresponding parameter. Second, notice that the standard error of the mean is estimated from sample values. Why do we estimate it? Because normal theory specifies that the actual standard error of the mean is given by $\dfrac{\sigma_x}{\sqrt{n}}$, but we rarely know what σ_x (the population standard deviation) is, so we estimate it from sample data. This helps us understand why our multiplier is a *t* score rather than a *z* score. The *t* distribution—which is really a family of normal distributions (see Fig. 8.3)—is used to estimate the population mean in situations when we have small-sample data and don't know σ. The various *t*

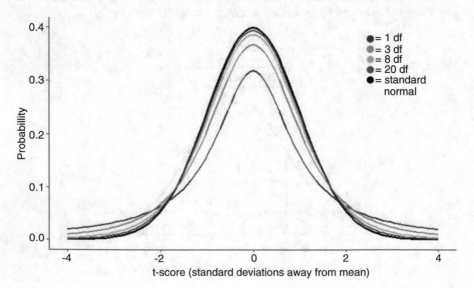

Fig. 8.3 Some *t* distributions for various sample sizes and the standard normal (*z*) distribution. (Source: Bevans, B. (November, 2020). https://www.scribbr.com/statistics/t-distribution/)

Table 8.1 Multipliers from the z and t distributions for several sample sizes

	Multiplier for a 95% CI
z (large n assumed)	± 1.96
t (n = 30, df = 29)	± 2.05
t (n = 20, df = 19)	± 2.09
t (n = 10, df = 9)	± 2.26

distributions are indexed by their degrees of freedom (df), which is simply n-1 in this example. So, when n is very large, the t and standard normal (or z) distributions are nearly identical and offer similar multipliers. As sample size gets smaller, however, the t distributions increasingly reflect more variability (because small-n estimates vary more). This means that the t multiplier will be larger than the z multiplier for a given confidence level. Table 8.1 presents some examples of multipliers from the z and t distributions for forming a 95% confidence interval. The larger multipliers for confidence intervals based on small samples produce less precise estimates, as we will see shortly.

The normal-theory confidence interval using a t multiplier is widely used for parameter estimation. And for good reason: its estimation accuracy (i.e., does it contain the parameter at the expected confidence level?) is very good, if two conditions are met: the sample is large (say, n > 100) *and* come from a random sample of the population. Even when those conditions are met, the normal-theory estimation method has a big limitation: it only works if you have a formula for calculating the standard error of the statistic from sample data. Researchers commonly use means in their research and therefore want to estimate the population mean. Conveniently, there is a formula for calculating the estimated standard error of the mean (see above). But many effect size statistics that we might want to use in research do not have a standard error that can be calculated from a formula. In the next section, we learn how bootstrapping extends the usefulness of normal-theory estimation methods and opens up alternative estimation methods that are not bound to normal-theory assumptions of having data from large random samples.

8.4 Bootstrapped Confidence Intervals

In this section we introduce two confidence interval methods that use bootstrapping: the *t interval with bootstrap standard error* and the *bootstrap percentile interval*. These two methods are the hammer and screwdriver of the confidence interval toolbox; they're intuitive, easily calculated, and accurate in most circumstances. Let's learn the R code for bootstrapping and how we use that to construct confidence intervals.

The *t interval with bootstrap standard error* follows the formula covered in the previous section (reproduced below) with one exception.

$$X \pm \left(SE_b * t_{.95} \right)$$

We substitute the bootstrapped standard error of the mean (SE_b) for the normal-theory standard error of the mean ($SE_{\bar{x}}$). This substitution allows the t interval to be used to estimate parameters whose standard error is not possible to calculate with a formula. For this first example, we'll stick with the sample mean and construct a 95% confidence interval to estimate the population mean using a bootstrapped, rather than the normal-theory generated, standard error of the mean. For this example, we revisit the Chap. 1 example of describing the weight of NHANES survey participants. Let's also imagine that our study has a sample size of 200, which we create below by randomly sampling 200 cases from the 2011–2012 NHANES survey data.

```
library(NHANES)
```

```
#create small dataset by selecting survey year and sampling from those cases
dat<-NHANES[NHANES$SurveyYr=="2011_12",]
library(dplyr)
```

```
new<-sample_n(dat, 200)
```

A bootstrapped standard error (SE_b) is simply the standard deviation of a distribution of bootstrapped statistics. Since we are interested in the bootstrapped standard error of the mean, we create a distribution of means from bootstrapped samples. Here's a run-through of the resampling operation needed to do that, explaining each step of the R code below.

First we need to define two objects.

- N is the number of bootstrap samples we want to draw. How do we set N? The overriding concern is that our bootstrapped estimate of the standard error is reliable—meaning that we want to be confident that independent bootstrapping procedures using the same data produce very close estimates, so that in turn we have confidence in our particular estimate of SE_b. The experts recommend that N = 1000 is adequate for generating bootstrap distributions with accurate estimates of the standard error. There's no cost to setting N higher (e.g., 2000), given the power of standard desktop computers, but Ns > 1000 won't return too much more accuracy.
- We set up a container to hold the N means this bootstrapping operation will produce and give it a name. In the example below, that container is called *boot*, but it's just a name and could be called anything.

Second, we set up a for-loop to organize the resampling operation: i stands for a particular bootstrap sample, and the operation inside the loop (defined by these brackets: {}) is done for each of our N samples. The for-loop below has two operations.

- First we randomly sample n (which in this case is 200) values (with replacement) from the weight variable. Those n values are stored in an object called *samp*. You can think of this as one bootstrap sample.
- Next we find the mean of the bootstrap sample created in the first line and pass that mean to the numeric container set up earlier.

The loop then repeats that two-step process for all N bootstrap samples. Once we have run the loop and created N means from bootstrapped samples, we can generate a histogram of the bootstrap distribution. Although not necessary for interval estimation, the histogram helps us visualize what the true population sampling distribution of the mean looks like. Finally, it is essential that we save the standard deviation of the bootstrapped distribution, which is the bootstrapped standard error (SE_b), in an object for later use.

```
#bootstrap distribution of the sample mean
N=1000
boot=numeric(N)
for (i in 1:N) {
  samp <- sample(new$Weight,200,replace=T)
  boot[i] <- mean(samp)
}
hist(boot, main="bootstrapped distribution of
NHANES respondent weight",xlab = "mean weight (kg) of n=200 sample")
```

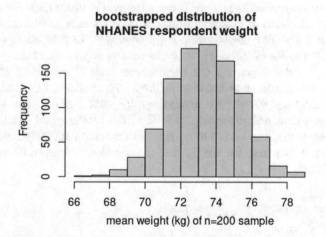

Now let's use R to find the *t* interval with 95% confidence to estimate the mean weight in the population (μ) of Americans. First we find the sample mean and store it in an object called *xbar*. Having found SE_b above, we find the appropriate *t*-score multiplier for 95% confidence with the qt() function and use those two quantities to calculate the margin of error (stored in an object called *moe*). Then we subtract the MOE from, and add it to, the sample mean for the lower and upper limits of the CI.

```
SEb=sd(boot)
SEb
```

```
## [1] 1.904111
```

```
#95% t with bootstrapped std error CI
```

```
xbar=mean(new$Weight)
n=200
moe=qt(0.975,n-1)*SEb
xbar-moe
```

```
## [1] 69.59868
```

```
xbar+moe
```

```
## [1] 77.10832
```

```
#or do in one operation
xbar+c(-1,1)*moe
```

```
## [1] 69.59868 77.10832
```

The *t* with bootstrap standard error estimation method produces a symmetrical interval: the range of plausible values of the parameter is the same below and above the statistic, or point estimate. This reflects the shape of a true population sampling distribution that is assumed to be normal. Accordingly, the *t* with bootstrap standard error interval works best when the sample data (and, presumably, the population) are approximately normally distributed. We will learn what "works best" means later.

The *bootstrap percentile interval* follows from an intuitive idea. For a 95% confidence interval, the most plausible values of the parameter are the middle 95% of the values of the bootstrapped distribution. Conversely, the least likely values of the parameter are the 5% of the values in the lower and upper tails of the bootstrapped distribution. By that logic, we can construct an interval consisting of percentile values from the bootstrapped distribution. For a 95% confidence interval, we simply identify the 2.5th and 97.5th percentile values. Regardless of the shape of the distribution, these values define the middle 95% of the bootstrapped statistics. All we need to generate this interval is a bootstrapped distribution of the statistic, which we already have. From that, we simply find the appropriate percentile values using quantile() function.

```
#95% percentile CI
quantile(boot,c(0.025,0.975),na.rm=TRUE)
```

```
##     2.5%    97.5%
## 69.54240 76.95824
```

These two interval estimation methods produced very similar intervals. That won't always be the case, but when it is, either interval is acceptable to report as your parameter estimate. We'll get to interpreting a confidence interval after one more example.

8.5   Estimation with Effect Size Statistics

Confidence intervals for estimating population means (as in the example above) are very common in research. That is in part due to the fact that the estimated standard error of the mean is easy to calculate from sample statistics. However, as we learned in Chaps. 3, 4, 5, and 6, research questions are generally addressed with effect size statistics, many of which don't have a formula for calculating their standard error. The bootstrapped confidence intervals we learned above can be used to estimate parameters associated with any effect size statistic, even if its standard error can be calculated from a formula.

In this example we continue using our n = 200 sample of 2011–2012 NHANES survey data. Our research question concerns the difference between females' and males' age of first sexual experience (see the NHANES documentation for the variable SexAge). We would like to use our sample data to estimate the difference in the population. Recall from Chap. 3 that several statistics could be used to describe the X-Y relationship; in this example we use the mean difference (MD). The chunk of output below shows the following, in order:

- A frequency table of our X variable, so that we can incorporate those group sizes into the resampling operation.
- Two objects, set up as in the previous example, where N is the number of bootstrap samples we want to draw from our data and md is a container to hold N numeric values, which in this case will be mean difference (MD) statistics.
- A for-loop that samples randomly with replacement from each group of our X variable, maintaining the respective groups' sample sizes from the original sample data. With these bootstrap samples of female and male data, we calculate the mean difference in age of first sexual experience and deposit that value in the numeric container.
- A histogram of the bootstrap distribution of sample mean differences in age of first sexual experience.
- SE_b is saved in an object called SEb.
- Code for calculating the 95% t and percentile confidence intervals and the intervals themselves.

```
table(new$Gender)

##
## female    male
##     89     111

#bootstrap distribution of the sample mean difference
N=1000
md=numeric(N)
for (i in 1:N) {
  gp1 <- sample(new$SexAge[new$Gender == "female"],89,replace=T)
  gp2 <- sample(new$SexAge[new$Gender == "male"],111,replace=T)
  md[i] <- mean(gp1,na.rm=TRUE)-mean(gp2,na.rm=TRUE)
}

hist(md, main="bootstrapped distribution of
     sample mean differences",xlab="years")
```

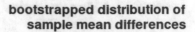

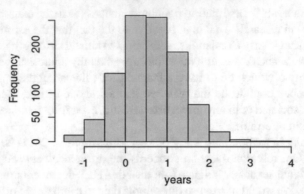

```
SEb=sd(md)

#use R to find group means and calculate MD
meandiff<-mean(new$SexAge[new$Gender=="female"],na.rm=T)-
  mean(new$SexAge[new$Gender=="male"],na.rm=T)
meandiff

## [1] 0.6015152

#t interval

n=200
moe=qt(0.975,n-1)*SEb
meandiff+c(-1,1)*moe

## [1] -0.8023265  2.0053568

#percentile interval
quantile(md,c(0.025,0.975))

##      2.5%     97.5%
## -0.757827  2.059731
```

Notice that the sample mean difference (MD = 0.602) is about 0.6 years, with females reporting an older age of first sexual experience than males. However, that is a sample statistic—what is the *true* (i.e., in the population of all Americans) mean difference? Our *t* confidence interval shows us that the parameter could range from females' first sexual experience being 0.76 years younger than males to 2.06 years older than males. The percentile confidence interval generates a very similar range of possible values of the parameter. We now turn to the subtleties of correctly interpreting a confidence interval estimate.

8.6   How to Interpret a Confidence Interval

A confidence interval consists of a range of plausible values of the (unknown) parameter—a range that is defined by lower and upper values—to which we attach a level of confidence. Confidence intervals are often interpreted with phrasing like

"we are 95% confident that the interval 'contains' or 'includes' the parameter."
However, a proper understanding of a confidence interval requires that you recognize that your particular confidence interval is just one of many that would have arisen from hypothetical bootstrapping procedures from the same data and yours could have been any one of those other intervals. To help visualize that, imagine an experiment in which random samples (of size n) are drawn from a population with a known mean and a confidence interval for estimating the population mean generated from each sample. The results from such an experiment are displayed in Fig. 8.4.

What do we learn about confidence intervals from this display? First, sample means (green dots), and in turn confidence intervals, vary randomly. Second, a few means vary so much from μ that their intervals don't even include the parameter (those shown in red). Indeed, for a 95% confidence interval, we expect that over a long series of random sample-generated intervals, 95% of them will *cover* the parameter (i.e., include it within the interval range) and 5% will not. That's what is shown above; out of 100 random samples, confidence intervals from 5 do not cover the parameter. Third, most sample means cluster closely around μ (2.25) and so across many hypothetical samples there will emerge some consensus on most plausible values of μ. In other words, not all values within a particular confidence interval are equally likely values of the parameter.

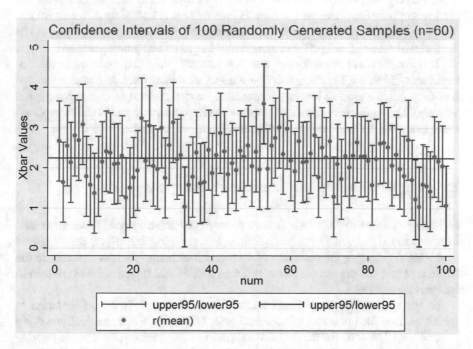

Fig. 8.4 Confidence intervals of the mean from random samples from a population with μ = 2.25.
(Source: https://sites.nicholas.duke.edu/statsreview/671-2/)

With those principles in mind, let's take another shot at confidence interval interpretation. Confidence in parameter estimation is indeed a long-term proposition. It really doesn't make any sense to say we have 95% confidence in a *particular* interval. When we say that we "have 95% confidence in" an interval of values containing the parameter, we mean that 95/100 intervals generated from hypothetical samples like ours will include the parameter. That interpretation, however, is not easy to render into a simple sentence, which is one reason why simpler (but somewhat inaccurate) interpretations gain currency.

Earlier we learned two bootstrap methods and used them to estimate a parameter: a t interval and a percentile interval. Having introduced the concept of confidence interval *coverage* above, let's now address the question of how we evaluate these confidence interval methods. An accurate interval is one whose coverage equals its confidence and whose "misses" are the same at both ends of the interval. For example, a 95% confidence interval should cover the parameter in exactly 95% of the samples from a simulation study (like the one illustrated in Fig. 8.4) and have 2.5% error rates on each side (estimating too high and too low at the same rate). Researchers have concluded that both the t interval and percentile intervals perform well in large samples. In small samples the t interval performs better, but both estimation methods have a narrowness bias, meaning that the true coverage is less than 95%. If the bootstrap distribution is not normally distributed, the percentile interval performs better.

Revisiting the example above in which we estimated the median math score in the kindergarten boys, the t and percentile intervals provide slightly different values. We have large N data, but given that the bootstrap distribution of medians is not normally distributed, the percentile interval is the more accurate estimator.

It is also worth reminding ourselves that although statistical conclusion validity indicates evidence for *some* non-random effect or relationship, we cannot assume that the predictor variable is solely responsible for that effect (i.e., causal conclusion validity). That inference, of course, depends on the control over alternative explanations that were designed into the study that produced the data.

8.7   Factors That Affect Confidence Intervals

In this section we explore two factors that affect bootstrap confidence intervals: sample size (n) and confidence level. Remember, anything that affects the margin of error (MOE) changes the precision of an interval estimate. Factors that reduce the MOE lead to more precise estimates, whereas factors that increase the MOE lead to less precise estimates.

To see how sample size affects a confidence interval, let's do a simulation in which we find the bootstrapped standard error (SE_b) while systematically varying sample size. The code for the first simulation (n = 100) is shown. A normal population was set up to sample from, with mean = 5 and standard deviation = 2. Then we simply applied the loop used in the example in Sect. 8.4 to generate a bootstrapped distribution of means for a particular sample size and find the standard deviation of that distribution. You can run it, as well as the other three simulations, but remember that your results will differ randomly from mine because you are doing independent sampling from the population data. The results of all four simulations are presented in Table 8.2—what do they show? First, notice the inverse relationship between n and SE_b. Larger samples generate means (or whatever statistic you're using to estimate the parameter) that vary less, which in turn yields smaller values of SE_b and smaller MOEs. Sample size is also related to the t multiplier used in the t interval, which has an additional effect on the MOE: larger n is related to smaller t values, which in turn yield smaller MOEs. Larger sample sizes also produce more precise percentile confidence intervals by the same logic. Means that vary less will result in bootstrapped distributions in which the 2.5th and 97.5th percentile values are closer together.

```
dat=rnorm(10000, 5, 2)    #set up a normal population with mu=5, sigma=2
N=1000
n=100
boot=numeric(N)
for (i in 1:N) {
  bootsamp <- sample(dat,n,replace=T)
boot[i] <- mean(bootsamp)
}
#bootstrapped standard error of mean
SEb=sd(boot)
SEb
```

```
## [1] 0.1982948
```

Thus far in this chapter, we have discussed 95% confidence intervals, but in fact the confidence level attached to an interval estimate is the analyst's or researcher's decision. A 95% confidence level is a common convention, but other values may be preferred depending on how the interval is used. For example, imagine that we have a treatment whose effectiveness in the population is estimated by an interval. A value of 0 would indicate that the treatment is not effective, and if 0 is a plausible value (is included in the interval estimate), then the treatment would not be offered to the public. However, if the consequences of offering the treatment are high (say

Table 8.2 Bootstrapped standard errors of the mean and margin of error for several sample sizes from a normal population

n	SE_b	t multiplier for 95% confidence (df = n-1)	MOE
100	0.20	1.98	0.40
50	0.28	2.01	0.56
25	0.40	2.06	0.83
10	0.63	2.26	1.43

the treatment has some nasty side effects), then we would want to be very confident that the treatment is indeed effective—so that's a situation for setting confidence high, perhaps 99%. On the other hand, if the consequences of *not* offering the treatment are high (say some people could die without it), then we might not need to be so confident about the treatment's effectiveness because a treatment that might not be highly effective is better than nothing—so that's a scenario for setting confidence lower, perhaps 85% or 90%. Those are considerations for setting confidence level when intervals are used to make decisions with practical consequences. When intervals are used simply to estimate a parameter, 95% is an acceptable all-purpose confidence level.

You can easily see how the confidence level affects interval precision. In a *t* interval with bootstrap standard error, the confidence level determines the *t* multiplier. Higher confidence levels produce larger *t* values, which in turn translate to less precise intervals. For example, referring to the table above (and you can explore this in R with the qt() function), if we wanted to generate a *t* interval with 99% confidence, the *t* multiplier (for n = 100/df = 99) would be 2.63. With a bootstrap percentile interval, the same result occurs: higher confidence levels produce quantiles that are more extreme, leading to less precise intervals.

8.8   Writing Up a Parameter Estimation Study

We will do much more with interval estimation in data analysis in Chap. 9, producing bootstrapped confidence intervals to estimate population effect sizes in studies with various statistical models. To end this chapter, here is a written summary of the example described earlier (Sect. 8.4).

Results

The relationship between gender and age of first sexual experience was estimated from a sample of NHANES 2011–2012 survey data (n = 200). The sample data showed that females (M = 18.1 years) were older than males (M = 16.8 years) at the time of their first sexual experience than males. Two estimation methods were used to estimate the population mean difference with 95% confidence. The t interval with bootstrap standard error (95% CI: −0.80, 2.0) and the bootstrap percentile interval (95% CI: −0.76, 2.1) produced similar estimates. In other words, the difference in age of first sexual experience is likely to be between 0.8 years (with females' younger) and 2.0 years (with males younger). A mean difference of 0 is a plausible value of the parameter, suggesting that females and males do not differ in the reported age of their first sexual experience.

8.9 Problems

Use the Melanoma data (205 patients with malignant melanoma) in the MASS package for problems 1–3.

1. Generate and interpret a bootstrapped percentile confidence interval (with 95% confidence) for estimating the mean survival time in the population of patients.
2. Generate separate bootstrapped percentile confidence intervals (with 95% confidence) for estimating the mean survival time in the populations of female (sex = 0) and male (sex = 1) patients, respectively. Interpret each. What do these confidence intervals say about the survival times of women and men?
3. Generate separate t with bootstrapped standard error confidence intervals (with 95% confidence) for estimating the mean survival time in the respective populations of patients with (ulcer = 1) and without (ulcer = 0) an ulcer. Interpret each. What do these confidence intervals say about the relationship between having an ulcer and melanoma survival time?
4. In the cats (MASS) data, estimate the mean body weight in the population of cats with 95% confidence intervals, using both t and percentile methods. Interpret the intervals. Explain the difference between the methods.
5. Generate separate percentile confidence intervals (with 95% confidence) for estimating the body weight in the respective populations of male (sex = M) and female (sex = F) cats. Interpret each. What do these confidence intervals say about the relationship between sex and cat body weight?
6. Using the birthwt (MASS) data, estimate the median birthweight in the population from which the sample of mothers came using a t with bootstrapped standard error confidence intervals (with 95% confidence). Then, generate separate t confidence intervals (with 95% confidence) for estimating the median birthweight in the respective populations of mothers who smoked (smoke = 1) and didn't smoke (smoke = 0) during pregnancy. Interpret each. What do these confidence intervals say about the relationship between mothers' smoking status during pregnancy and the infant birthweight?

Chapter 9
Using Resampling Methods for Statistical Inference: Four Examples

Covered in This Chapter
- How to do systematic data analysis
- Examples of randomization for hypothesis testing and bootstrapping for parameter estimation in four statistical models

9.1 Introduction

This chapter pulls together the statistics appropriate for exploratory and descriptive data analysis (covered in Chaps. 3, 4, 5, and 6) with the resampling methods used for hypothesis testing (covered in Chap. 7) and effect size estimation (covered in Chap. 8). This chapter also pulls together the rationale, actions, and disciplines that comprise good data analysis, and that have been applied in the data analytic examples throughout this book, and renders them into a seven-step process. The goals of this chapter are to present four data analytic examples that model the seven-step data analytic process and incorporate the descriptive and inferential statistical methods we have learned thus far. As in previous chapters, this chapter will also address the written communication of results from in a way that reflects the systematic data analytic process presented below.

9.2 Doing Systematic Data Analysis: A Seven-Step Process

Assume two things: one, you have been given a file of data or a way to access a data file, and, two, you (or someone you're working with) have a question of interest that they believe the data will address. What do you do next? How do you ensure that your data analysis is as unbiased as possible (i.e., won't just find what you're

© The Author(s), under exclusive license to Springer Nature Switzerland AG 2023
B. Blaine, *Introductory Applied Statistics*,
https://doi.org/10.1007/978-3-031-27741-2_9

Table 9.1 Steps and tasks in the data analytic process

Step		Tasks
1	Data and documentation	File format issues
		Source of documentation
2	Read in and explore data frame	Data object in R
		Variable types and dimensions
		Missing data analysis
3	Identify variables and statistical model	Variables needed to address research question
		Statistical model
		Sampling and study design issues
4	Explore/describe/summarize	Statistical lens(es) for analysis
		Statistical and graphical summaries
		Explore influential values
5	Significance testing	Assumption of exchangeability
		Randomization test method
6	Parameter estimation	Bootstrapping method
		Confidence interval type
		Confidence level
7	Communication	Level and needs of end user
		Data and data analytic record

looking for)? How do you ensure that you don't miss something potentially important in the data that you weren't looking for? And how do you make sure that your data analytic operations are done so that they're transparent (i.e., there's no mystery as to what you did) and replicable (i.e., someone else should be able to redo your analysis)? Below (see Table 9.1) I lay out the steps in a data analytic process that guides you through "what to do next?" in a systematic way, so that your analysis:

- Addresses the research question.
- Balances exploratory and confirmatory data analytic goals.
- Controls bias.
- Maintains transparency and replicability in your work.

Some of the tasks raised in the steps below go beyond the teaching objectives of this book, or what the examples in this book demonstrate. We mention them nevertheless because they are important elements to the data analytic enterprise and instructors may want to address those.

Step 1. Data and Documentation

The first step in data analysis is to access the original data file and determine the shape and format of the original data. Assuming the data file is in some spreadsheet form, a *wide* data file consists of cases or observational units (e.g., people) in rows and variables in columns, and a particular case's data is only in one row. A *long* data file formats the data from each case over two or more rows. For example, data for a particular year might be formatted over 12 rows, each of which has monthly observations. Often, data that is entered and managed in spreadsheets is in long format, but for statistical analysis, a wide format is recommended. Therefore, step 1 may involve reshaping the data file. It is common to enter variable names in row 1 of a spreadsheet or data file, and that is another detail to confirm before reading the data into R. Finally, it is important to know how variables in the original study were measured, how or if they were coded for data entry, and what their names indicate. Hence, data documentation, or *meta-data*, is generally essential to the data analyst. As we have seen throughout this book using data files that are part of R packages, documentation is indispensable for working with variables as well as for interpreting statistics.

Step 2. Read in and Explore Data Frame
Once the issues in step 1 have been addressed, you must get the data into R for analysis. Commonly, data files are "read in" with R functions designed to read the type of data in their particular format (e.g., .txt, .csv). What we want to do is create a data object for working with in R. It is important for transparency and replicability to not make changes to the original data file prior to bringing it into R. Changes that you might want to make to the original data file, such as recoding a variable or creating a new variable from a combination of others, should be done as part of your analysis, so there's a record of those changes in your code.

After getting the data in the form of a data frame in R, it is recommended to explore the dimensions and properties of that data object. Some examples of this exploration are numbers of rows and columns, the variable type (e.g., fac, int) assigned to each variable by R, and simple statistical summaries of each variable. Another task in this step is to size up the missing data problem. Missing data is common, but it is good to assess whether the missing data are "missing at random" or whether there is a nonrandom pattern of missing data (e.g., survey participants from a particular subgroup didn't answer questions). Although this book has not covered methods for detecting and analyzing missing data, it is an important task, and your instructor may add some material to address it.

Step 3. Identify Variables and Statistical Model
In step 3 we turn to the research question and determine what variables in the data file are being used in the question, which is X and Y, and how those variables are measured. If research questions are posed very generally (e.g., What is the relationship between studying and academic achievement?"), multiple variables in the data file might be relevant to each variable, in which case the analyst must consult with the person asking the research question, or make their own decisions about the variables being used in the analysis and why. Once the X and Y variables are set, the statistical model underlying the research question becomes an explicit guide to the statistics and plots generated as part of the analysis. It is part of understanding the statistical model to also diagnose the sampling method and design in the original study, if that is possible. Recall from Chap. 2 that those elements affect the interpretation of statistics, which in turn affect how one communicates the findings (step 7).

Step 4. Explore/Describe/Summarize
Step 4 involves the descriptive analysis, using statistics and plots, that allows us to summarize the relationship in the research question. We also should do some

exploratory data analysis, in which (generally through plots) we explore the data for characteristics and patterns that could inform or qualify the relationship summarized in effect size statistics. For example, an overall relationship might be very small, but exploration of the data reveals that there are distinct subgroups (e.g., female vs male participants) in the data with very different relationships to the outcome variable. As the examples Chaps. 3, 4, 5, and 6 demonstrated, exploratory analysis also allows us to detect possible influential (or even incorrect) values and make some decisions about how to handle those values in subsequent analyses.

Step 5. Significance Testing

Once we have described the X-Y relationship in the research question with one or more effect size statistics, we need to determine if the size of that relationship is statistically significant. Step 5 focuses on null hypothesis significance testing, which we have learned to do with randomization methods. The task in this step is to set up the test that reflects the statistical model (i.e., what statistic is being used?) as well as the design elements of the original study (i.e., is the assumption of exchangeability met?). If two statistics were used to summarize the relationship, such as the mean difference and Cohen's d in an ANOVA model study, then the analyst conducts randomization tests of the null hypothesis associated with each statistic. Of course, the p-value resulting from the test will need to be interpreted and reported.

Step 6. Parameter Estimation

As with significance testing, step 6 addresses another important inferential task: estimating the population effect size of the X-Y relationship being examined. The main task here is to set up and run the bootstrapping procedure for the desired statistic. The analyst may choose to generate an interval estimate of the parameter with a level of confidence that is less (for more precision) or greater (for more confidence) than 95%. The analyst can pick from a range of theoretically different confidence intervals as well, although we covered only the two most common in this book. Once the interval is generated, it becomes part of the communicated findings.

Step 7. Communication

In step 7 the analyst is largely concerned with two tasks. First, regardless of whether the data analytic findings will be communicated in a report, presentation, slideshow, or talk, the analyst must decide (or help the researcher decide) how to best communicate with the end users of the findings. This has implications for both how much statistical detail to report and how much jargon and technical language to use. Second, the analyst must preserve the data analytic record, as well as the original data file, so that the various decisions made in the analysis are transparent and the analysis can be replicated. This is done by saving the R script file (which, ideally, is detailed with annotations) and the data file in such a way that they can be shared if requested by other parties.

The next section features four full data analytic examples. Chiefly these are examples of how to do null hypothesis testing with randomization tests and parameter estimation with bootstrapping using effect size statistics from the four statistical models that were covered in Chaps. 3, 4, 5, and 6. However, the examples also illustrate the seven-step data analytic process described above. For each example, we do the following, in order:

(a) Describe the steps in the analysis in a series of bullet points, including notes about data analytic decisions and R functions.
(b) Conduct the statistical computing and generate the output in R.
(c) Present a write-up of the analysis.

9.3 Data Analysis in an ANOVA Model: An Example

With this seven-step data analytic process to work with, let's do an example with data from an ANOVA model. In Chap. 3 we described effect of treatment on remission time (measured in weeks) in data from a sample of leukemia patients who were randomly assigned to a treatment or control condition. You might want to look back at the example in Chap. 3 (Sect. 3.6) and pull up the data documentation to refresh your memory; those will help you address the tasks in steps 1 and 3 of the data analytic process. Let's continue with that example, adding the statistical inference tasks for hypothesis testing and parameter estimation.

The chunk of output below shows the following, in order:

- Missing data analysis is part of step 2 above, and we do this with two R functions we haven't used before. When you pass your data frame object to the is.na() function, R returns a matrix of "true" (if there's missing data in a cell) and "false" (if there's data in a cell) statements. While the matrix of statements is not too helpful, we can use it to generate missing data totals for each variable, by counting the "trues." In this way we can see if a variable in our analysis is affected by missing data. We do that by passing the is.na() function to the ColSums() function. The output shows no cells in our data frame with missing data.
- Step 4 analysis is next and is centered around the statistical lens being used to examine the research question, which in this example is the mean difference. Histograms of remission time by group graphically summarize the relationship. We find the mean difference in remission time, storing it in an object for later use. Notice that the mean difference is calculated to be positive strictly for interpretive convenience. Finally, we use the boxplot methods demonstrated in Chaps. 1 and 3 to identify any remission time outliers, of which there are none.
- In step 5 we set up a randomization test of the null hypothesis. Although we rarely report formal hypothesis statements in a written summary, we should know what statement we are testing, so here are the null and alternative hypotheses for this example:

  H_o: Treatment (compared to control) has no effect on remission time.
  H_a: Treatment (compared to control) has an effect on remission time.

- To be precise, we are conducting a Monte Carlo randomization test, since we are sampling from the population of all possible random shufflings of 42 values into

2 groups of 21, and since the assumption of exchangeability holds in the data (i.e., patients were randomly assigned to treatment vs control in the original study). As a result, the randomization test below generates an estimated, not an exact, p-value. The procedure generated a distribution of the mean difference under H_o, pictured in the histogram. The estimated p-value ($p = 0.0024$)[1] of finding a mean difference in remission time of 8.4 weeks or greater, if H_o is true, constitutes strong evidence against the null.

- Step 6 calls for parameter estimation (i.e., the likely treatment effect on remission time in the *population* of leukemia patients), and this is done by bootstrapping the two confidence intervals we learned the previous chapter. The bootstrapping methods from Chap. 8 are used below. Notice that we randomly sample (with replacement) survival time from each experimental group to create a bootstrapped sample, then find the mean difference from that sample (called a bootstrapped mean difference), and then repeat 1000 times. We then apply the methods for finding each confidence interval from Chap. 8. Since the resulting confidence intervals are very similar, only one need be reported.

```
##ch9 example
#missing data
colSums(is.na(gehan))

##  pair  time  cens treat
##    0     0     0     0

#explore/describe
library(lattice)
histogram(~time|treat,data=gehan,
        type="count",
        breaks=seq(0,40,by=2.5),
        layout=c(1,2))
```

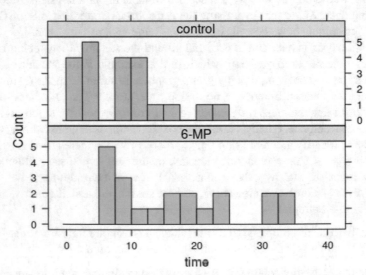

---

[1] Remember that you and I are doing independent computing, so your randomization test will be (randomly) different than the one below, with a randomly different p-value too.

```
MD<-mean(gehan$time[gehan$treat=="6-MP"])-mean(gehan$time[gehan$treat=="control"])
MD
```

```
## [1] 8.428571
```

```
#influential values
b<-boxplot(time~treat,plot=F,data=gehan)
b$out
```

```
## numeric(0)
```

```
#randomization test of Ho
#test statistic
obs=MD<-mean(gehan$time[gehan$treat=="6-MP"])-
  mean(gehan$time[gehan$treat=="control"])

N=5000
md=numeric(N)
for (i in 1:N) {
  data <- sample(gehan$time,42, replace=FALSE)
  md[i] <- mean(data[1:21])-mean(data[22:42])
}
hist(md, main="reference distribution of treatment effect
     on survival time under Ho",xlab="mean difference (weeks)")
```

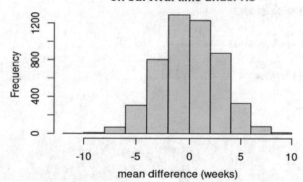

```
pvalue=length(which(abs(md)>=obs))/N
pvalue
```

```
## [1] 0.0024
```

```
#parameter estimate
N=1000
md=numeric(N)
for (i in 1:N) {
  gp1 <- sample(gehan$time[gehan$treat == "6-MP"],21,replace=T)
  gp2 <- sample(gehan$time[gehan$treat == "control"],21,replace=T)
  md[i] <- mean(gp1)-mean(gp2)
}

hist(md, main="bootstrapped distribution of
     treatment effect on survival time",xlab="weeks")
```

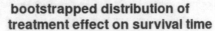

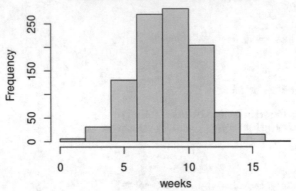

```
SEb=sd(md)

#t interval
n=42
moe=qt(0.975,n-1)*SEb
obs+c(-1,1)*moe

## [1]   3.289578 13.567565

#percentile interval
quantile(md,c(0.025,0.975))

##       2.5%      97.5%
## 3.569048 13.476190
```

Results

The effect of treatment on remission time (weeks) in a convenience sample (N = 42) of leukemia patients was investigated in an experiment in which participants were randomly assigned to 6-MP or a control condition. Means and standard deviations for remission time are in Box Table 9.1, and histograms of remission time are displayed in Box Fig. 9.1.

Patients in the 6-MP condition had longer remission time than control condition patients by an average of 8.4 weeks. A Monte Carlo test showed that difference was statistically significant, with an estimated p-value (p = 0.0024) indicating strong evidence against the null hypothesis. A bootstrapped 95% confidence interval (3.3, 13.6) provides a range of plausible values of the true treatment effect on remission time, which range from about 3–13 weeks.

Box Table 9.1 Remission time statistics by treatment group

Condition	n	Mean	SD
6-MP	21	17.1	10.0
Control	21	8.7	6.5

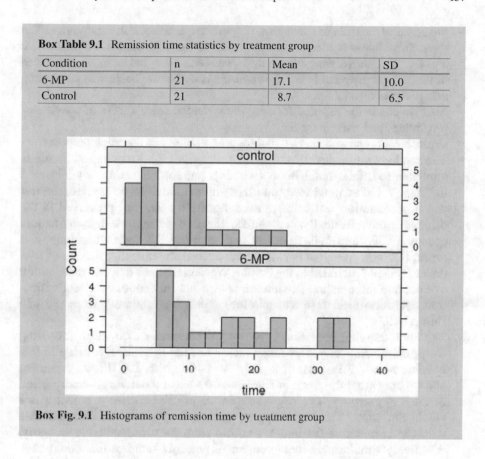

Box Fig. 9.1 Histograms of remission time by treatment group

9.4 Data Analysis in a Proportions Model: An Example

For this example we return to the Chap. 4 example using data from a 5-year study of breast cancer patients (see Sect. 4.6). As we did earlier, we assume that the study sample is a convenience sample, and we analyze the relationship between menopausal status (X) and disease remission (Y) using the risk ratio as the summary descriptive statistic.

The chunk of output below shows the following, in order:

- Create new factor variables (with new names and category labels) from the 0,1 variables in the gbsg data. This is done mainly so that our tables and plots will be more readable.

- We generate the table of conditional proportions, keeping X in columns and Y in rows. This table is needed for two purposes: to compute the RR and to generate a plot of the data. Missing data analysis was done using the methods introduced in the first example but is not shown in the code below. There was no missing data in our two variables.
- Next we create the function for computing the RR from a table of conditional proportions and retrieve the RR.
- Create an appropriately labeled plot for presentation and reporting purposes.
- The randomization test of the null hypothesis (H_o: Menopausal status is unrelated to illness status in breast cancer patients) is set up as we learned in Chap. 7 and adapted for proportions model data. To be precise, the test was a permutation test because exchangeability was not preserved in the original study. Basically, we shuffle the 686 patients into two random groups of 396 and 290, calculate and save the RR for that sample, and repeat N times. Note that the histogram of the reference distribution of RRs under H_o is not included in the output. We nevertheless use that distribution to calculate the p-value. [Reminder: if you run this code, your randomization operation (and resultant p-value) will vary randomly from the one shown below.]
- The bootstrapping operation is set up for parameter estimation following the principles in Chap. 8. Notice that we use the original Y variable in this operation—why? Because it is a 0,1 variable (with 1 = illness presence) and we know that the mean of a series of 0 s and 1 s delivers the proportion of the "1" category. The new variable we created earlier is a factor and would not work. The bootstrapping operation generates random samples with replacement of 0 s and 1 s within each menopause status category, effectively simulating a large number of possible samples that could have been in the original study. For each we calculate and save the RR. The resultant distribution of RRs, then, reflects the likely *true* (i.e., in the population) relationship between menopausal status and illness status. Finally, we use this distribution of RRs to generate 95% confidence intervals for estimating that parameter.
- As step 7 calls for communication and reporting of the analysis and results, we generate a short write-up of the analysis.

```
library(survival)
gbsg$menopause.status=factor(gbsg$meno,level=c(1,0),labels=c("post","pre"))
gbsg$illness.status =factor(gbsg$status,level=c(1,0),labels=c("recurrence/death","no
 recurrence"))

#frequency and proportions tables
tab<-table(gbsg$illness.status,gbsg$menopause.status)
t<-prop.table(tab,margin = 2)
t

##
##                        post        pre
##    recurrence/death 0.4545455 0.4103448
##    no recurrence    0.5454545 0.5896552

#find RR
RR<-function(tab){
  rr=tab[1,1]/tab[1,2]
  print(rr)
}
RR(t)

## [1] 1.107716

#barplot
barplot(t,
        ylim=c(0,1),
        beside=T,
        legend=rownames(tab),args.legend=list(x="topleft"),
        main = "Menopause status and 5-year survival
        of breast cancer patients")
```

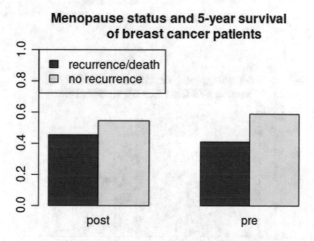

```
#randomization test of Ho
table(gbsg$menopause.status)
##
## post  pre
##  396  290

obs<-RR(t)

## [1] 1.107716

N=5000
riskratio=numeric(N)
for (i in 1:N) {
  data <- sample(0:1,686, replace=T)
  grp1 <- data[1:396]
  grp2 <- data[397:686]
  riskratio[i] <- mean(grp1)/mean(grp2)
}
hist(riskratio,main="reference distribution of effect of
     menopausal status on illness status under Ho", xlab="RR")

pvalue=length(which(riskratio>=obs))/N
pvalue

## [1] 0.0922

#parameter estimate

N=1000
rr=numeric(N)
for (i in 1:N) {
  grp1 <- sample(gbsg$status[gbsg$menopause.status=="post"],396,replace=T)
  grp2 <- sample(gbsg$status[gbsg$menopause.status=="pre"],290,replace=T)
  rr[i] <- mean(grp1)/mean(grp2)
}
hist(rr,main="bootstrapped distribution of effect of
menopausal status on illness status",xlab="RR")
```

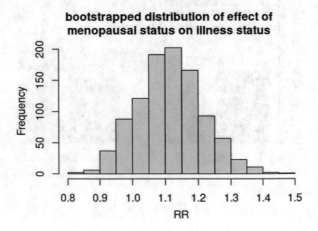

```
SEb=sd(rr)
#t interval
n=686
moe=qt(0.975,n-1)*SEb
obs+c(-1,1)*moe

## [1] 0.9116366 1.3037950

quantile(rr,c(0.025,0.975))

##      2.5%      97.5%
## 0.9369207 1.3182835
```

Results

Breast cancer patients (N = 686) who were either premenopausal or post-menopausal were followed for 5 years. The outcome variable was illness status (cancer recurrence/death or not) at the end of the trial. The proportions of patients by group in both illness status categories are displayed in Box Fig. 9.2. The cancer recurrence or death proportion was slightly higher in postmenopausal (0.45) than premenopausal patients (0.41). The risk ratio (RR = 1.11) shows that cancer recurrence or death was 1.11 times higher in postmenopausal compared to premenopausal patients, which is a small relationship. A permutation test showed very weak evidence against the null hypothesis (p = 0.092). A bootstrapped 95% confidence interval (0.91, 1.3) for estimating the true relationship includes 1.0 as a possible value of the parameter, which reflects the weak evidence from the permutation test. We cannot conclude that menopausal status is related to illness status.

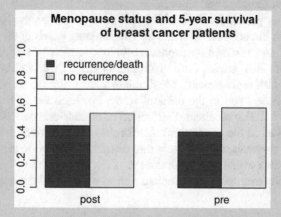

Box Fig. 9.2 Proportion of pre- and postmenopausal breast cancer patients with cancer recurrence or death after 5 years

9.5   Data Analysis in a Regression Model: An Example

For this example we return to the Chap. 5 example (see Sect. 5.6) using data on atmospheric conditions in New York City from 1973 (which is part of the survival package) and continue our analysis of the relationship between daily high temperature (X, in degrees Fahrenheit) and average daily ozone concentration (Y, in parts per billion).

The chunk of output below shows the following, in order:

- The missing data analysis reveals that there are no missing data cells in the ozone and temperature variables.
- The descriptive analysis consists of two tasks, which were the focus of the Chap. 5 example. One, generate a scatterplot of the X-Y relationship with the least-squares regression line plotted on it, including our write-up. Two, fit the model and retrieve the model fit statistics and regression coefficient.
- The randomization (technically a permutation) test follows the principles that we learned in Chap. 7, and in this example tests the null hypothesis that daily high temperature is unrelated to ozone concentration (or, H_o: $\beta_1 = 0.00$). Notice that when we call for the coefficients from a regression model we get both β_0 and β_1. When we set up the randomization test, however, we only want β_1. The [2] in the creation of the object that holds our observed statistic tells R to use the second value in that list of coefficients. The logic of the resampling operation below is as follows: create random samples of ozone and temperature data by sampling with replacement from the original variables, fit the regression model in that sample, retrieve β_1 and store it in a numeric vector, and then repeat N times. That procedure produces a distribution of regression coefficients under the null hypothesis, which naturally is centered on $\beta_1 = 0$. The p-value in the analysis below is 0, but only because there were no occurrences of a regression coefficient greater than 2.44, the observed value.
- The bootstrapping procedure differs in that it must recognize the X-Y relationship that exists in the original data, and that presumably exists in the population. For that reason the resampling procedure below randomly samples pairs of observations (i.e., X-Y values for a given participant). By sampling rows from the data frame (with replacement), we simulate a large number of possible samples that could have been in the original study. For each we fit the regression model and save β1. The resultant distribution of regression coefficients reflects the likely true (i.e., in the population) relationship between daily high temperature and ozone concentration. We use this distribution to generate 95% confidence intervals for estimating the parameter.
- As in our previous examples, we generate a short write-up of the analysis.

```
library(survival)
library(lattice)

#missing data
colSums(is.na(environmental))

##        ozone   radiation temperature      wind
##            0           0           0         0

#scatterplot, model and coefficient statistics
xyplot(ozone ~ temperature, data=environmental,
       grid=TRUE,
       type=c("p", "r"),      #p=scatterplot, r=regression line
       col.line="red")
```

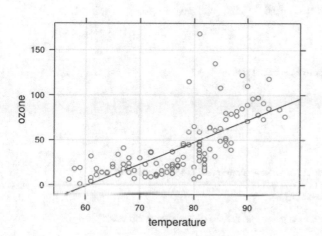

```
reg=lm(ozone ~ temperature, data=environmental)
reg$coefficients

## (Intercept) temperature
##   -147.64607     2.43911

summary(reg)$sigma

## [1] 23.92025

summary(reg)$r.squared

## [1] 0.4879601

#randomization test of Ho
beta<-reg$coefficients[2] #retrieve beta1 only

N=5000
rcoeff=numeric(N)
for (i in 1:N) {
  y=sample(environmental$ozone,111,replace=F)
  x=sample(environmental$temperature,111,replace=F)
  mod<-lm(y~x)
  rcoeff[i] <- mod$coefficients[2]
}
hist(rcoeff,main="regression coefficient distribution under Ho")
```

regression coefficient distribution under Ho

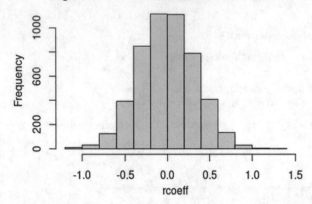

```
pvalue=length(which(abs(rcoeff)>=beta))/N
pvalue

## [1] 0

#parameter estimate
N=1000
boot=numeric(N)
for (i in 1:N) {
  dat <- environmental[sample(nrow(environmental),111,replace=T),]
  mod <- lm(ozone~temperature,data=dat)
  boot[i] <- mod$coefficients[2]
}
hist(boot,main="bootstrapped regression coefficient distribution")
```

bootstrapped regression coefficient distribution

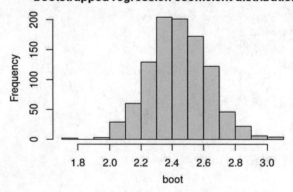

```
#t interval
SEb=sd(boot)
n=111
moe=qt(0.975,n-1)*SEb
beta+c(-1,1)*moe
```

```
## [1] 2.071426 2.806793
```

```
#percentile interval
quantile(boot,c(0.025,0.975))
```

```
##     2.5%    97.5%
## 2.084544 2.815961
```

Results

The relationship between daily high temperature (°F) and ozone concentration (ppb) in New York City was examined in data from May to September of 1973 (N = 111). A scatterplot of the scores is in Box Fig. 9.3. The least squares regression coefficient (β = 2.44) showed ozone concentration increased about 2.4 ppb for each degree increase in daily high temperature. Daily high temperature explained about 49% of the variability in ozone concentration (R^2 = 0.49). A randomization test showed that the relationship between daily high temperature and ozone concentration was statistically significant and there was strong evidence against the null hypothesis ($p < 0.001$). A bootstrapped 95% confidence interval (2.1, 2.8) suggests that the population relationship likely falls between 2.1 and 2.8 ppb increase in ozone concentration for each additional degree of daily high temperature.

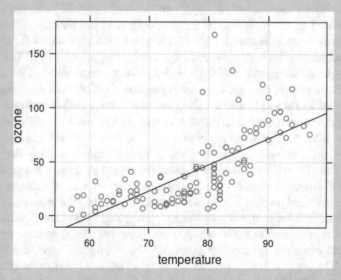

Box Fig. 9.3 Daily high temperature and ozone concentration in New York City (May through September 1973)

9.6   Data Analysis in a Logistic Model: An Example

To demonstrate the data analytic process and statistical inference operation in logistic model data, we return to example two from Chap. 6 (Sect. 6.8). Recall, we are using data from the German Breast Cancer Group study (gbsg), a trial of 686 patients with breast cancer, to address the relationship between number of positive cancer nodes (X) on illness status (Y, cancer recurrence or death before the end of the trial: yes/no).

The chunk of output below shows the following, in order:

- The missing data analysis reveals that there are no missing data cells in the status and nodes variables.
- The descriptive analysis consists of two tasks. First, we need to fit the logistic model and generate the statistics needed to describe the X-Y relationship. This is done below with the logistic regression coefficient, R^2_{logistic}, and the point-biserial correlation coefficient. Second, we need to generate a plot to both visualize the relationship and to include in the write-up. A barplot requires a table of conditional proportions, which in turn requires that we decide on a way to split the numeric predictor into groups for plotting. As we demonstrated in the Chap. 6 example, exploring the range of the nodes variable reveals that three groups are probably best. The output below also generates the proportions table, for our own use and in case we want to report any of those values. Notice that we label the 0,1 status variable with category labels for tabling and plotting purposes only, changing its name in the process. Below, in the inferential portion of the analysis, we must use the original numeric 0,1 variable.
- The randomization (again, technically a permutation) test follows the principles that we learned in Chap. 7, and in this example tests the null hypothesis that number of cancer nodes is unrelated to illness status (or, H_o: $\beta_1 = 1.0$). Remember, in logistic regression an odds ratio of 1.00 indicates "no relationship." The structure and logic of the randomization procedure reflect what we did in the regression model example above. We save the regression coefficient in the form of an odds ratio (again using [2] when calling the coefficients list). The logic of the resampling operation below is as follows: create random samples of illness status and cancer nodes data by sampling with replacement from the original variables, fit the regression model in that sample, retrieve β_1 and convert the log-odds to an OR, store that OR in a numeric vector, and then repeat N times. That procedure produces a distribution of odds ratio coefficients under the null hypothesis, which naturally is centered on $\beta_1 = 1.0$. This histogram displaying that reference distribution is not included in the output below. Notice that to find the p-value from this distribution, we must know whether our observed OR is greater or lower than 1.0 and enter the appropriate operation ($\geq$, in this case) into the p-value statement. The p-value in the analysis below is 0, but only because the reference distribution produced by the randomization test did not include any ORs greater than 1.11, the observed value.

- The bootstrapping procedure logic is also similar to that done in a regression model. The resampling procedure below randomly samples pairs of observations (i.e., *X-Y* values for a given participant), to preserve the relationship between nodes and survival status. For each we fit the logistic regression model and save β_1. The resultant bootstrapped distribution of ORs reflects the likely true (i.e., in the population) relationship between cancer nodes and survival status. We use this distribution to generate 95% confidence intervals for estimating the parameter.
- Finally, we generate a short write-up of the analysis.

```
#missing data
colSums(is.na(gbsg))
##           pid            age          meno          size         grade
##             0              0             0             0             0
##         nodes            pgr            er        hormon       rfstime
##             0              0             0             0             0
##        status illness.status        nodes3
##             0              0             0
```

```
#fit model and retrieve statistics
reg=glm(status ~ nodes,data=gbsg,family=binomial)
exp(reg$coefficients)
```

```
## (Intercept)       nodes
##   0.4573587   1.1126478
```

```
Rsq<-((reg$null.deviance) (reg$deviance))/reg$null.deviance
Rsq
```

```
## [1] 0.04657132
```

```
cor.test(gbsg$nodes,gbsg$status)$estimate
```

```
##       cor
## 0.2422873
```

```
#proportions table and barplot
gbsg$illness.status=factor(gbsg$status,level=c(1,0),labels=c("recurrence/death","no
 recurrence"))
library(Hmisc)
gbsg$nodes3=cut2(gbsg$nodes,g=3)
tab=table(gbsg$illness.status,as.factor(gbsg$nodes3))
p=prop.table(tab,margin=2)
p
```

```
##
##                     [1, 3)    [3, 6)    [6,51]
##   recurrence/death 0.3131313 0.4350282 0.6084906
##   no recurrence    0.6868687 0.5649718 0.3915094
```

```
barplot(p,col=c("blue","red"),
        ylim=c(0,1),beside=T,
        legend=rownames(tab),args.legend=list(x="topright"),
        main="Proportion of cancer recurrence/death
        by number of cancer nodes")
```

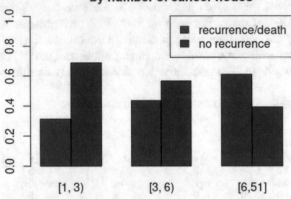

```
 #randomization test of Ho
OR=exp(reg$coefficients[2])
N=5000
rcoeff=numeric(N)
for (i in 1:N) {
  y=sample(gbsg$status,686,replace=F)
  x=sample(gbsg$nodes,686,replace=F)
  mod<-glm(y~x,family=binomial)
  rcoeff[i] <- exp(mod$coefficients[2])
}
hist(rcoeff,main="odds ratio distribution under Ho")

pvalue=length(which(rcoeff>=OR))/N
pvalue

## [1] 0

#parameter estimation
N=1000
boot=numeric(N)
for (i in 1:N) {
  dat <- gbsg[sample(nrow(gbsg),686,replace=T),]
  mod <- glm(status~nodes,data=dat,family=binomial)
  boot[i] <- exp(mod$coefficients[2])
}
hist(boot,main="bootstrapped distribution for
     estimating population OR")
```

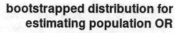

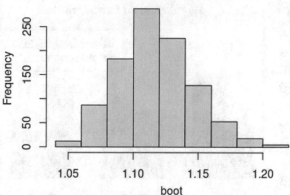

```
  #t interval
SEb=sd(boot)
n=686
moe=qt(0.975,n-1)*SEb
OR+c(-1,1)*moe

## [1] 1.056616 1.168680

#percentile interval
quantile(boot,c(0.05,0.95))

##       5%       95%
## 1.071816 1.166373
```

Results
The relationship between number of positive cancer nodes and cancer recurrence (yes/no) in a sample of breast cancer patients (N = 686) was examined with a logistic regression analysis. The regression coefficient (OR = 1.11) showed that each additional node was associated with 11% higher odds of cancer recurrence compared with no recurrence. The OR and the correlation coefficient (r_{PB} = 0.24) indicate a small positive relationship between nodes and recurrence. Patients' number of positive nodes explained about 5% of the variability in recurrence. A randomization test showed that the relationship between number of cancer nodes and illness status was statistically significant and there was strong evidence against the null hypothesis (p < 0.001). A bootstrapped 95% confidence interval (1.06, 1.17) suggests that in the population, each additional cancer node is associated with somewhere between 6% and 17% greater odds of recurrence or death by the end of the trial (Box Fig. 9.4).

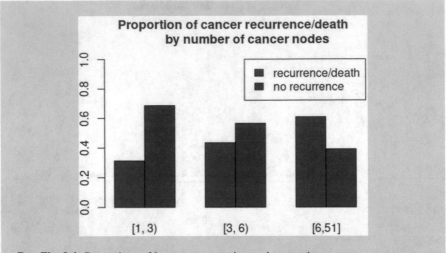

Box Fig. 9.4 Proportions of breast cancer patients who experience cancer recurrence or death as a function of the number of cancer nodes

9.7   Statistical Significance and Practical Significance

We have learned that resampling methods—randomization tests and bootstrapping—are powerful and flexible tools for addressing two important statistical inference questions:

- Is there any effect of predictor on outcome (or relationship between predictor and outcome) beyond what we would expect to happen randomly, given that samples and sample statistics vary?
- What is the probable size and direction of the effect (or relationship) in the population?

Both of these tasks use sample data as a surrogate population, resampling from it to generate a distribution of outcomes that allows us to make probabilistic inferences about the questions above. Resampled distributions approximate the population outcome distributions very well if the original sample is reasonably representative of the population.

It is important that both inference tasks are part of the data analytic process introduced at the beginning of this chapter, because they help us distinguish between the *statistical significance* and the *practical significance* of a finding. In Chap. 7 we learned that "statistical significance" is the term used to refer to relationships in which there is some evidence against H_o. A common practice in social and health sciences research is to use a threshold p-value to declare statistical significance, and p < 0.05 is a conventional threshold. That's another way of say-ing "if there's at least weak evidence against H_o, we will call this finding statisti-cally significant." We emphasized in Chap. 7 that a p-value is not an effect size statistic, and it bears repeating here. A statistically significant X-Y relationship does not mean that the X-Y relationship is large, important, meaningful, or use-ful—or in other words, *practically significant*. A randomization test of the null hypothesis can comment on statistical significance but not practical significance. A bootstrapped estimate of the parameter, however, allows the analyst to com-ment on practical significance. Whereas the assessment of statistical significance is done with a p-value, the assessment of practical significance often requires additional knowledge or context. For example, prior studies on the research ques-tion and how their effects are interpreted can provide context for what experts in the field regard as an important. An appreciation of the units of outcome measure-ment, and what they mean to practitioners in the field, is another clue to practical significance.

One final comment about distinguishing statistical and practical significance needs to be mentioned. In a null hypothesis test, an observed effect or relationship is essentially compared to no effect (e.g., a mean difference of zero). This often ends up being a "straw man" test, because the null hypothesis (i.e., an intervention having *zero* effect on the outcome) is an unlikely outcome to begin with, and so finding evidence against H_o is often not too surprising. The reality in many fields is that research needs to establish that a new intervention or treatment is more effective than current practice—not more effective than zero. In this way confi-dence interval estimates of a treatment effect or X-Y relationship actually help the researcher assess both statistical and practical significance. For example, say a study finds that a new drug (compared to placebo) lowered cholesterol by 10 points (95% CI: 5, 15). Let's imagine that the currently used drug has a benefit of about 8 points. So, the confidence intervals tell us that our study finding has sta-tistical significance (i.e., the null value, 0, is not in the interval so is not a plausible value of the parameter). But the CI also includes the value of the current drug's effect (8 points), which means that the new drug's effect cannot be distinguished from the current treatment's effect. In this way confidence intervals allow the analyst and researcher to test *non-null* hypotheses (e.g., Is a new treatment better than the current treatment?).

9.8   Problems

The problems below all use datasets in the MASS package. Use `library(MASS)`
to make the datasets available and? `datasetname` *to see the documentation for*
each, which will provide variable names and measurement details.

1. In the `cats` dataset, test the relationship between `Sex` (x) and the `Bwt` (y) with
   a randomization test of the null hypothesis using the mean difference. Find and
   interpret the p-value.
2. Using the `cats` dataset, estimate the population difference between male and
   female cats' bodyweight with a bootstrapped 95% confidence interval.
3. In the `Melanoma` dataset, summarize the relationship between survival time
   (days) of males and females with the mean difference. Test the null hypothesis
   and generate a 95% confidence interval estimate of the parameter. Summarize
   your findings.
4. How do your statistical inferences in #3 change if you use the median difference
   to summarize the relationship between sex and survival time? Rerun your
   randomization and bootstrapping procedures with the median difference and
   summarize the findings.
5. In the `Melanoma` dataset, test the null hypothesis that `sex` (x) and `ulcer` (y)
   are unrelated using a risk ratio. If you want to generate a summary table of conditional proportions, create a new factor variable with labels from the original
   numeric outcome variable for that purpose (see example below), but use the raw
   0,1 variable in your randomization test.

```
Melanoma$ulcer_f=factor(Melanoma$ulcer,level=c(0,1),labels=c("presence","
absence"))
```

6. In the `Melanoma` data, estimate the relationship between sex and ulcer in the
   population with a 95% confidence interval estimate. Generate and interpret a t
   with bootstrapped standard error CI and a percentile CI.
7. In the UScereal dataset, do a randomization test of the relationship between
   sugars (X) and the calories (Y) data using the regression coefficient as
   your statistical lens. Find and interpret the p-value. For a challenge, rerun
   the test to test the null hypothesis that the correlation coefficient (Pearson
   r) is 0.
8. Is the relationship between mother's weight (X) and infant birth weight (Y)
   statistically significant? Using the data in the `birthwt` data, run a randomization test of H_o and interpret the p-value? Does a statistically significant X-Y
   relationship mean that the relationship is large? Explain?
9. Using the regression coefficient, estimate the population relationship
   between mother's weight and infant birthweight in the `birthwt` data,

using a 95% confidence interval. When doing parameter estimation with a regression coefficient, what does it mean if $\beta = 0.00$ is in the confidence interval?

10. In the `birthwt`, test the null hypothesis associated with the relationship between number of previous premature labors (X, ptl) and whether the baby is low birthweight (Y, coded $0 = $ not low, $1 = $ low). Set up and run a randomization test using the OR. Interpret the p-value with regard to H_o. Next, estimate the population relationship with bootstrapped 95% confidence intervals. Generate both the t and percentile intervals and interpret. Summarize your findings.

Chapter 10
Statistics and Data Analysis in a Pre-Post Design

Covered in This Chapter
- Descriptive statistics for summarizing pre-post data
- Hypothesis testing and parameter estimation with pre-post data
- Internal validity in pre-post designs

10.1 Introduction

In each of the statistical models covered earlier in this book (Chaps. 3, 4, 5, and 6), participants were measured *once* on the outcome variable. In ANOVA and proportions models, participants were in only one category of the predictor (their X value) and provided a score on the outcome (their Y value). In the regression and logistic models, each participant provided a pair of observations—one X value and one Y value. A very common study type among researchers takes the form of a *repeated measures model*. A repeated measures model is one in which participants provide more than one value on the outcome variable and therefore are measured at two or more points in time on the outcome variable. Although repeated measures models can accommodate categorical outcomes, numeric outcome variables are much more common.

$$\text{numeric } Y = \beta_0 + \beta_1 \text{time } X + \varepsilon$$

Notice that in a repeated measures model *time* is the predictor variable. As with all of our previous statistics, we want to describe the X-Y relationship, which in this case is how time (i.e., being measured at two or more points in time) is related to the outcome. In many repeated measures studies, the treatment or intervention is *coincident with* the time predictor. For example, we might investigate the effect of mental imaging (motivational vs. cognitive imaging) on athletic performance (basketball

© The Author(s), under exclusive license to Springer Nature Switzerland AG 2023
B. Blaine, *Introductory Applied Statistics*,
https://doi.org/10.1007/978-3-031-27741-2_10

foul shot accuracy) by having a sample of athletes practice the motivational imagery treatment and take the foul shot performance test at time 1 and then at time 2 (which could be minutes, hours, or days later) practice the cognitive imagery treatment and take the foul shot performance test. Notice that the sample of athletes is in both treatment conditions but, relevant to this chapter's topics, gets measured twice on the outcome variable. We would be interested in comparing those two sets of scores for evidence of some relationship between imagery type and performance.

Other repeated measures model studies impose the treatment *between* the repeated measures. One such study type that is widely used among researchers is the *pre-post design*. In a pre-post study, the researcher examines the *X-Y* relationship by observing participants before and after the treatment or intervention. This generates pretest (before the intervention) and posttest (after the intervention) observations from the same participants, which can be compared to analyze the effect of the intervention. Pre-post designs are intuitive and appealing because they allow researchers to observe how a group of participants change as a result of some intervention. Many research questions can take this form, such as:

- Does taking an SAT prep course improve test scores?
- How does smoking behavior change before and after a smoking cessation workshop?
- Do athletes' performance improve after mental imagery training?

Repeated measures studies allow us to summarize and test the effect of time on some outcome variable. It is important to recognize that that "effect" of time in a repeated measures study includes the treatment or intervention but also includes other possible explanations of the change in the outcome from time 1 to time 2. We will discuss this design limitation, and its implications for how effects from repeated measures model studies are interpreted, in a later section. For now let's turn to the descriptive statistics and plots used to summarize *X-Y* relationships in pre-post data.

10.2   Data Analysis in Pre-Post Data

Imagine we are interested in the effect of a new method for teaching fractions on third grade students' understanding of fractions. We did a study which measured understanding of fractions in a class of ten students with a 10-point test on Monday. Then, the teacher delivered four lessons based on the new teaching method throughout the week. On Friday the students took the test again. In this example, the intervention is the new teaching method consisting of four lessons, and the outcome—the understanding of fractions—is measured before (pretest) and after (posttest) the intervention. It would be easy to see this in ANOVA model terms, as we did in Chap. 3. We do, in fact, have two groups of scores on a numeric outcome variable, and the mean difference in a logical effect size statistic.

In an ANOVA model, each participant contributes only one outcome score, so when we compare group means, we are comparing scores from two different groups of participants. Using a signal-to-noise ratio analogy, the treatment effect in an

ANOVA model study is the signal. The noise consists of all the ways the individuals in those two groups differ—in their physical and psychological traits, past experiences, skills and ability, and more. To "see" the signal (i.e., the effect of the treatment) in an ANOVA model study, the treatment effect must overcome the noise of the total random differences in the two groups' participants. The fact that participants contribute a pair of outcome observations in a pre-post study means that we are able to see how *individuals* change, rather than merely groups of individuals. And because the same participants are in both "groups," there is much less noise in a pre-post study, which make it easier to observe a treatment effect (if there is one). Pairs of outcome observations (i.e., within an individual) therefore are a stronger statistical lens for measuring the effect of an intervention or treatment that occurs between those observations.

So, rather than treating the pretest and posttest means as if they were from different groups of students, the better statistical approach is to transform the students' two scores into one score by taking the difference their pretest and posttest scores. These transformed scores are called *difference scores* (symbolized with D). The differences between pretest and posttest scores can also be called *gain scores* and *change scores*. For the rest of the chapter, we will use the term *change scores* because the observational unit in a pre-post study is the individual's change over time. "Gain" scores imply a direction of change that may be confusing if the point of the intervention is to lower scores on the outcome. Also, using change scores avoids the confusion with the mean difference (MD) statistic used in ANOVA model studies. Pretest, posttest, and change scores for our hypothetical study of knowledge of fractions are shown in the plot in Fig. 10.1 (the dark squares show the pretest and posttest means, respectively) and in Table 10.1.

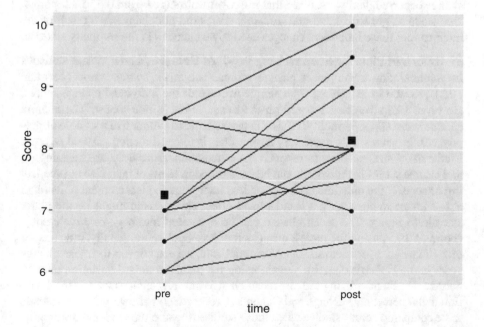

Fig. 10.1 Change scores from a hypothetical pre-post study

Table 10.1 Pretest, posttest, and change score for a hypothetical study of 10 students' understanding of fractions

Participant	Pre	Post	D
1	7	9	2
2	6.5	8	1.5
3	6	6.5	0.5
4	8	7	−1
5	8	8	0
6	6	8	2
7	8.5	8	−0.5
8	7	9.5	2.5
9	8.5	10	1.5
10	7	7.5	0.5

The change scores (still symbolized with D) from our hypothetical study were calculated by subtracting the pretest from the posttest scores, for each participant. This is often the order of calculation when the intervention is presumed to increase scores on the outcome, because positive change scores indicate an improvement on the outcome from pretest to posttest. The order of calculation (e.g., pre minus post, or post minus pre) is arbitrary and the analyst's decision, but it is important to know what the sign of the D scores mean in terms of the pre-to-post change. As with a correlation coefficient, the sign and magnitude of the pre-post change are important properties of the D statistic. In this example, positive D scores indicate improvement in knowledge of fractions, and it appears that 7/10 (or 70%) of the students made some improvement. The size of the D scores indicates how much (or little) change occurred, and we can see that improvement ranged from 0.5 to 2.5 points. Two students got worse and one student's scores did not change as a result of the intervention; those individual changes need to be included in the summary statistic.

Statistics and Plots for Summarizing Pre-Post Data Let's now look at statistics for summarizing a sample of change scores. The *mean change score* (Formula 10.1), referred to as "D-bar," is a simple average of the individual change scores. However, like all statistics based on the mean, D-bar is non-robust. The *median change score* (D_{med}, Formula 10.2) is a better statistic to summarize a sample of pre-post change scores if there are very large values in either direction. $\bar{D}$ and D_{med} are useful effect size statistics to report if the outcome measure units are meaningful, and change scores in those units can be understood in terms of their size or practical importance. If the outcome variable has less interpretable measurement units (e.g., a 1–7 Likert rating scale), a standardized effect size statistic might be the better statistic to report. The mean change can be converted into a *percentage change* (Formula 10.3) using the pretest mean as the "baseline" score. Finally, much as we did with mean differences in ANOVA model data, we can convert the mean change score into a *Cohen's d* statistic, again using the pretest standard deviation to standardize $\bar{D}$. Standardizing the raw mean change into percentage or standard deviation units (percentage change and Cohen's *d*, respectively) allows a change statistic to be compared across studies when those studies measure the outcome differently.

$$\bar{D} = \sum(D_i)/n \tag{10.1}$$

$$D_{med} = \text{median}(D_i) \tag{10.2}$$

$$\text{percentage change} = (\bar{D}/M_{pretest}) * 100 \tag{10.3}$$

$$\text{Cohen's } d = \frac{\bar{D}}{s_{pre}} \tag{10.4}$$

Let's compute and interpret these statistics in our example of students' understanding of fractions before and after a series of lessons. The mean and median change scores ($\bar{D} = 0.9$ and $D_{med} = 1.0$, respectively) indicate that student's scores improved an average of around one point on the 10-point test scale. In terms of percentage change, the mean change was a 12.4% increase in test scores over the pretest level of understanding. Finally, Cohen's $d = 0.95$ meaning that the change in students' scores was equivalent to about a 1 SD improvement from pretest to posttest.

Now let's consider graphical summaries of pre-post data and exploring the data for extreme values. Because we have converted students' pretest and posttest scores to *one* change score, we can summarize pre-post data with the same methods we learned in Chap. 1. By way of a quick review, a histogram allows us to see the whole distribution of change scores. A histogram displays location (i.e., typical change scores) and variability (i.e., the general spread of change scores) and is particularly good for identifying asymmetry. If change scores are skewed or distributed asymmetrically, D_{med} may be a better summary statistic than $\bar{D}$. Similarly, a density plot works with sample data to develop through resampling an estimate of the population distribution of change scores and could also be used to portray location and variability and diagnose asymmetry. A boxplot is an alternative graphical summary tool that is organized around the median, quartiles, and adjacent values of the distribution of change scores. Recall that a boxplot is also a diagnostic tool for identifying outliers, which may have disproportionate influence on summary statistics like $\bar{D}$ and s_{pre} (both of which are used in the calculation of Cohen's d).

Statistical Inference with Pre-Post Data Now that we have statistics and plots for summarizing change scores in a pre-post design, we should proceed to questions of statistical inference. In terms of our example, this would involve testing the null hypothesis that there is no effect of the treatment (i.e., lessons) on understanding of fractions (H_o: $\bar{D} = 0$). If we apply the principles and methods from Chap. 7, this would involve a randomization test of one of the summary statistics covered earlier, probably the mean change score. The principle of exchangeability is not met in our example's data, meaning that students were not randomly assigned to treatment conditions. In fact, there is only one treatment condition—the intervention that occurred between the pretest and posttest measurement—and all students were in that condition. The p-value generated by the randomization (which in this example is a permutation, Monte Carlo) test would be interpreted as evidence against the validity of the null hypothesis.

The internal validity of a pre-post study (see Chap. 2) must recognize that although we might speak of the "treatment effect," the predictor variable in a pre-post study is *time*. And although the treatment in our example (e.g., lessons on fractions) occurred in the time interval between pretest and posttest, so did lots of other things, at least hypothetically. And many of those things are also reasonable alternative explanations for the observed change in scores. The design elements of a pre-post study, and their implications for internal validity and causation, will be explored more in a later section. Generalizability, however, is still related primarily to the sampling method in the study. If the study data is from a random sample of the population of interest (i.e., the population of third grade students), the study findings have generalizability to the population. However, the particular features of a pre-post study add a layer of complexity to the assessment of generalizability. In repeated measures designs, participants are measured more than once over a period of time. It is possible—perhaps even likely in certain conditions—that some of those participants will "drop out." They will be included in the pretest measurement but for whatever reason(s) will not be measured at posttest. It is reasonable to wonder if such dropouts are somehow systematically different (e.g., they're not as interested in math) than the participants who remained in the study and therefore might subtly change the "population" the sample represents. This loss of data over repeated measures threatens both internal validity and generalizability actually, and we will explore more fully later in the chapter.

Finally, statistical inference includes parameter estimation, which would involve generating an estimate of the "true" (i.e., in the population of students from which our sample came) change in score from pretest to posttest from our sample data. If we apply the principles and methods from Chap. 7, this would involve generating a bootstrapped confidence interval estimate of the effect size of interest, which again would probably be $\bar{D}$ or Cohen's d but might be D_{med} if the sample data were very asymmetrically distributed. The confidence interval would give a range of plausible values of the parameter and show the range of outcomes we might have observed in our sample data had we used a different random sample.

10.3   Data Analytic Example 1

In this example we use data from a study of young female patients with anorexia, which is available in the anorexia dataset in the MASS package. The documentation shows that the patients were assigned to one of three treatment conditions, with pretreatment and post-treatment measures of weight (in lbs.). For this example we analyze the data from just those participants in the family treatment condition. Our research interest is whether, and how much, family treatment affects weight change in young females with anorexia.

The chunk of output below shows the following data analytic operations, in order:

- Address two data wrangling tasks: First, select the participants in the family treatment condition and assign their data to a new data object. Second, create a new variable consisting of the difference between post-treatment and pre-treatment weight. The particular order of calculation yields a change variable in which a positive value indicates weight gain.
- Generate the four effect size statistics covered in the previous section to summarize the effect of family treatment on weight.
- Generate plots to visualize the distribution of weight change values and explore the distribution for influential values via the boxplot rule. Due to the small sample, the bin width in the histogram must be fairly wide to achieve a reasonable level of smoothing.
- Run a randomization test of the null hypothesis. Given the design features of a pre-post study, the randomization test requires a new method to generate an appropriate probability distribution. Here's the basic logic of the method: Under the null hypothesis (assuming that family therapy has zero effect on weight), the pre-treatment and post-treatment measures of weight are simply two measures of weight from the same person separated only by time. Although the formal assumption of exchangeability we learned in Chap. 7 doesn't apply here because there is no random assignment of participants to condition, the idea is nevertheless useful. The two measures of weight, for any given participant, can be thought of as exchangeable. Because they're only randomly different, those measures are just as likely to have occurred in the opposite order. Given that either order of measures within participants is equally likely under H_o, and both ordering result in the same *size* difference, the difference between the two measures will differ only by *sign*.
- How is this logic translated to R code? Notice that the first step in the procedure is to create a vector of 1 s and −1 s by randomly sampling 17 values from a data object consisting of the integers 1 and −1. Then, that random vector of signs is applied to the change scores observed in the sample data, creating a vector of change scores under H_o. The third step is to find the mean of those random change scores, save that in a data object, and repeat N times. The loop generates a distribution of change scores (in lbs.) that simulates what is expected under the null hypothesis, with the now-familiar center at zero. The p-value is calculated with our standard method. Remember that if you run this randomization code yourself, your probability distribution and p-value will be randomly different from this example's output.
- To estimate the parameter (i.e., the mean weight change in the population of female patients with anorexia), we generate a bootstrap distribution of D-bar statistics in random samples of n = 17 from the population, using the sample data

as a surrogate population. The logic and computing method to do this is very similar to the method we used to bootstrap a regression coefficient. By randomly sampling rows of data from our data frame (i.e., pairs of observations, which in this case are pre-treatment and post-treatment weights), we create a bootstrap sample in which the relationship between treatment and weight is maintained. Next we calculate the D scores for the 17 "participants" in that sample. Then we calculate the mean of those D scores, which is our bootstrapped D-bar for one interaction. Saving those bootstrapped statistics over N iterations completes the loop. The resulting bootstrapped distribution is used to generate a confidence interval estimate, and our example shows both t and percentile 95% confidence intervals.

```
library(MASS)

#subset participants in the FT condition
new <- anorexia[anorexia$Treat=="FT",]
#create change variable
new$change <- new$Postwt-new$Prewt


#statistics
Dbar <- mean(new$change)
Dmed <- median(new$change)
percentchange <- Dbar/mean(new$Prewt)
cohensd <- Dbar/sd(new$Prewt)
Dbar

## [1] 7.264706

Dmed

## [1] 9

percentchange

## [1] 0.08728532

cohensd

## [1] 1.448107

#plots
hist(new$change,prob=T)
lines(density(new$change))
```

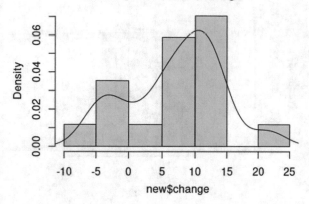

```
boxplot(new$change,horizontal = T)
```

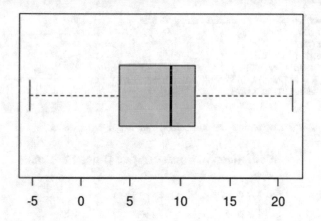

```
#randomization test of Ho

N=5000
diff=numeric(N)
for (i in 1:N) {
  signs=sample(c(1,-1),17,replace=T)
  samp=new$change*signs
  diff[i]=mean(samp)
}

hist(diff,main="probability distribution under Ho")
```

probability distribution under Ho

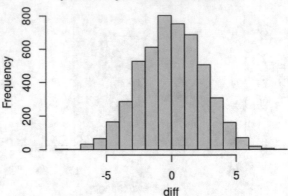

```
#find p value
pvalue=length(which(abs(diff)>=Dbar))/N
pvalue

## [1] 0.0012

#bootstrapped parameter estimate using D-bar
N=2000
boot=numeric(N)
for (i in 1:N) {
  df <- new[sample(nrow(new),17,replace=T),]
  df$change <- df$Postwt-df$Prewt
  boot[i] <- mean(df$change)
}

hist(boot,main="distribution of bootstrapped D-bar statistics")
```

distribution of bootstrapped D-bar statistics

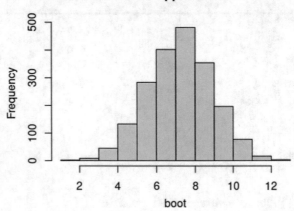

```
#t interval
SEb=sd(boot)
n=17
moe=qt(0.975,n-1)*SEb
Dhar+c(-1,1)*moe
## [1]  3.718647 10.810765

#percentile interval
quantile(boot,c(0.025,0.975))

##      2.5%     97.5%
##  3.952794 10.482353
```

Let's work through the interpretation of these data analytic results. Female anorexia patients in the study gained an average of 7.3 lbs. between pretest and post-test. Given the mean pretest weight of the patients as a baseline, this effect constituted an 8.7% gain in average weight. The median weight gain (D_{med} = 9 lbs.) was slightly larger, suggesting a somewhat negative skew in the distribution of D scores; this is noticeable in both plots. The average weight gain is equivalent to a gain of 1.45 standard deviations (Cohen's d = 1.45). All of these effect size statistics indicate that the family treatment occasioned a moderate to large effect on weight. The boxplot revealed no influential values, which is important to know because any outlier in such a small sample would have considerable influence on non-resistant statistics like the mean and standard deviation of change scores and, in turn, Cohen's d.

The randomization test returned strong evidence (p = 0.0012) that the observed mean change score (D = 7.26) was a very unlikely outcome under the null hypothesis. Can we attribute that change to the treatment and the treatment alone? Probably not, given that other factors could have occurred between the pre-treatment and post-treatment measurement and could explain at least some of that change. We will dive into those factors, and the reasons for the low internal validity of a pre-post design, in the next section. Finally, given that we have evidence of a non-random change in weight, our parameter estimate—estimated with 95% confidence—is that the "true" weight change (i.e., in the population of young females with anorexia) due to family treatment is between 3.9 lbs. and 10.5 lbs.

10.4 Internal Validity in Pre-Post Designs

In Chap. 2 we introduced two types of inference that, while not *statistical* in the strict sense of the word, are nevertheless part of what we do when we conduct, interpret, and report data analysis. The generalizability of the sample finding and whether there's a causal relationship between X and Y are both connected to the design of the study whose data we are analyzing. Generalizability is linked to the sampling method that generated the study data and applies to pre-post designs in the same way as every other statistical model we have covered: data from a random sample of the population (particularly if the sample is large) is associated with high generalizability of findings. This section focuses on internal validity of pre-post studies. To review, the internal validity of a study is the degree to which a study establishes that an observed change in *Y* is caused by *X* and *X* alone. Another way to say that is that studies with high internal validity have controlled all or nearly all confounding variables. Confounding variables, or "confounds," are variables that loom in a study and can explain at least some of the observed outcome. Remember, confounding variables don't have to be measured or empirically demonstrable to be a problem; they just need to be plausible. The internal validity of a pre-post study is threatened by any variable that *could* account for the observed change in scores. Let's look at some common types of variables that threaten the internal validity of, or "confound," pre-post studies and see how they loom as plausible alternative explanations.

Confounds Related to Time   In a pre-post design, the treatment variable occurs either coincident with the pretest and posttest measurement (e.g., mental imagery in athletes example) or between the two measurements (e.g., understanding fractions example). Internal validity requires that the observed change in scores is due to the treatment *alone*, but many other variables are operating within the time period between pretest and posttest measurement, and some of those variables could affect the outcome variable, and hence the change scores being analyzed. First, there are variables, called *history* confounds, that occur outside the study and could be related to the outcome variable measurement. If those events affect some or most of the study participants, the variable becomes a threat to the study's internal validity. For example, what if a bunch of the students in the fractions study just by chance were in the same after-school homework session and the tutor worked on fractions? That's a history confound because now we can't be sure if the improved score is due to the in-class lessons or to the after-school practice, or a mix of both. The threat due to history confounds increase with the time interval between pretest and posttest. Another time-related type of threat to internal validity concerns the changes in participants that naturally occur over time. If those changes are plausibly related to the outcome variable measurement, we have a *maturation* confound. Threats due to maturation are potentially bigger problems in studies that use children or adolescent participants, or have a long time (e.g., weeks or months) between pretest and posttest measurements, or both. In the fractions study, students would not be expected to mature or develop within a week in ways that could plausibly improve their scores.

Confounds Related to Repeated Measurement   As mentioned earlier a pre-post design allows researchers to observe change in participants, which is pretty compelling. However, repeated measurement of participants in a study also introduces several threats to internal validity. The first type is called a *testing* confound. This occurs when the experience of taking a test once affects how study participants take it the second time. All sorts of learning occurs when people are measured (especially if the measurement is of the self-report, paper-and-pencil variety) that can influence their responses on the test if it is taken again, such as understanding the vocabulary used on the test, getting used to the types of questions and responses required by the test, and more. So even in the absence of a treatment or intervention, study participants' scores on the posttest might differ from the pretest because of their familiarity with the testing instrument and procedure.

A pretest can also signal participants that they are in a study and give clues to what the study is about (e.g., its general aim or purpose). Even more consequentially, the pretest may heighten awareness in participants about what the treatment is supposed to do. These sorts of problems are called *testing-treatment interaction* confounds, because they identify ways that pretesting changes how participants prepare for, experience, and respond to the treatment, which, in turn, can change scores on the posttest and lower the internal validity of the study. In the fractions study, it is not unreasonable to imagine that at the point of the pretesting students knew there would be some lessons on fractions and another test after that. The teacher may have even said as much! Regardless, testing-treatment interaction effects loom in the study. Did students improve *solely* due to the new method lessons? Or did they improve in part because they were more motivated to improve their score, or wanted to impress the teacher, or any of the other ways that being aware of the purpose of the study could have helped them improve?

Third, there is a statistical tendency with repeated measurement or testing for people to score less extreme over time, a problem called *regression to the mean*. This is particularly worrisome in pre-post studies when participants are recruited or chosen for the study *because of* their extremely high (e.g., heavy smokers enrolled in a smoking cessation study) or low (e.g., poor readers enrolled in a study to improve reading) score. The statistical problem is that the extreme pretest score (whether it is low or high on the scale) is much more likely to get less extreme than more extreme, simply by chance. Regression to the mean always looms as an alternative explanation to the treatment effect in pre-post studies but is especially problematic when participants enter the study based on their low or high score on the outcome.

A fourth confound that arises with repeated measurement is when the measurement instrument itself (which includes the measurement medium and context) changes from pretest to posttest. Imagine that in our fractions study we measured understanding of fractions on Monday (pretest) with a paper-and-pencil quiz but the posttest measurement used a digital quiz taken online. This type of internal validity threat is called an *instrumentation* confound. Changes in the measurement of the outcome variable over time, even if they're subtle (e.g., a 5-point rating scale at pretest and 7-point at posttest), can affect change scores in ways that is not due to the intervening treatment variable.

Attrition   Attrition, or the loss of participants from a study, is an inescapable reality in research and confronts researchers working with all study designs and statistical models. Attrition becomes a threat to internal validity only when it becomes a plausible alternative explanation for the observed treatment effect. The principle works like this: when the lost participants are systematically different from the remaining participants, those two "groups" of participants now differ by a factor or factors in addition to the treatment variable. In a pre-post design, participant loss occurs over time and can result in posttest scores that reflect the influence of the lost participants. Imagine in our pretend study of third graders' knowledge of fractions, several of the students who "don't like math" skip the day on which the posttest is given. It is reasonable to think that those students might have been among the low scorers anyway, and so their absence inflates the posttest mean and produces a larger improvement than would have been observed had they remained in the study. As a potential threat to a study's internal validity, two things matter in diagnosing an attrition threat. First, the size of the participant loss, as a proportion of the original sample size, matters. The loss of a few participants from a large sample is expected and not problematic, even if they can be shown to be systematically different from the rest. Second, for an attrition threat to loom, the dimension(s) that the lost participants are believed to differ from the remaining participants must be plausibly related to the outcome. If, for example, the fractions study loses participants due to illness, that's less of a concern than if students select themselves out of the study. Illness would not be expected to remove students from a study in ways that bias the outcome measurement.

Let's sum up. As you can see, pre-post designs have low internal validity. As a result, they cannot establish that the treatment caused the observed outcome (i.e., the change in scores from pretest to posttest). The threats to validity covered above can be controlled or reduced with various study design elements and procedures (i.e., control condition, random assignment, avoiding a pretest). But those "fixes" would lead to a different statistical model. As long as pre-post designs remain popular and widely used, despite their limitations where establishing causation is concerned, we need to know how to analyze, interpret, and report their data. Our response should be to let our data analysis reporting communicate what the study can and cannot establish with regard to both generalizability and causation. Although our reporting may refer to the "effect" or "impact" of treatment in a pre-post study, we should avoid leaving the impression that the treatment had a causal effect on the outcome.

10.5 Writing Up a Descriptive Analysis

A written summary of the data analytic example in Sect. 10.3 is below.

Results
The effect of family treatment on weight (lbs.) in a sample of female anorexia patients (N = 17) was examined with a repeated measures design. Participants weighed, on average, 7.3 lbs. more at post-treatment compared to pre-treatment. Based on the participants' pre-treatment weight (M_{pre} = 83.6 lbs.), this gain represented an 8.7% average increase in weight. According to Cohen's d, the treatment effect was large (Cohen's d = 1.45). A randomization test showed that this weight change was statistically significant (p = 0.0012) with strong evidence against the null hypothesis. A bootstrapped 95% percentile confidence interval (4.0 lbs., 10.5 lbs.) estimated the effect of family therapy in the population (Box Fig. 10.1).

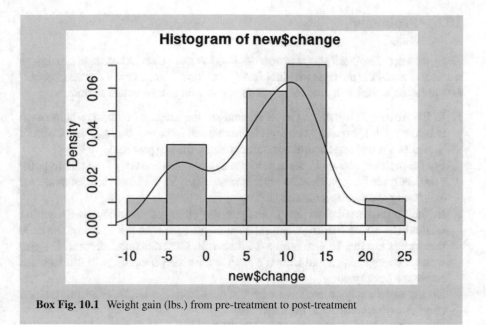

Box Fig. 10.1 Weight gain (lbs.) from pre-treatment to post-treatment

10.6   Problems

The problems below all use datasets in R packages. Use `library(package name)` *to make the datasets available and?* `datasetname` *to see the documentation for each, which will provide variable names and measurement details.*

1. In the `anorexia` dataset (MASS), analyze the effect of cognitive-behavioral treatment (CBT) on weight change. Generate and interpret $\bar{D}$, D_{med}, and Cohen's *d*. Analyze the change scores for outliers using the boxplot rule.
2. For the problem above (1), set up and run a randomization test of the null hypothesis. Interpret the p-value. Generate bootstrapped 95% confidence intervals for estimating the parameter and interpret.
3. In the `Baumann` dataset (car), analyze the effect of an innovative teaching method (Strat) on reading comprehension. Assume that the outcome variable measure is number of errors on a 16 item test. Generate and interpret $\bar{D}$, percentage change, D_{med}, and Cohen's *d*. Plot the change scores with a histogram and add a density curve.
4. Do the same analysis as above with the data from the DRTA teaching method condition. Summarize the findings.
5. For 3 above, run a randomization test of the null hypothesis and interpret the p-value. Generate bootstrapped 95% confidence intervals for estimating the parameter and interpret.

Printed in the United States
by Baker & Taylor Publisher Services